Cecil Balmond

Number 9

really fast.
Love, Mom
Merry Christmas
2010

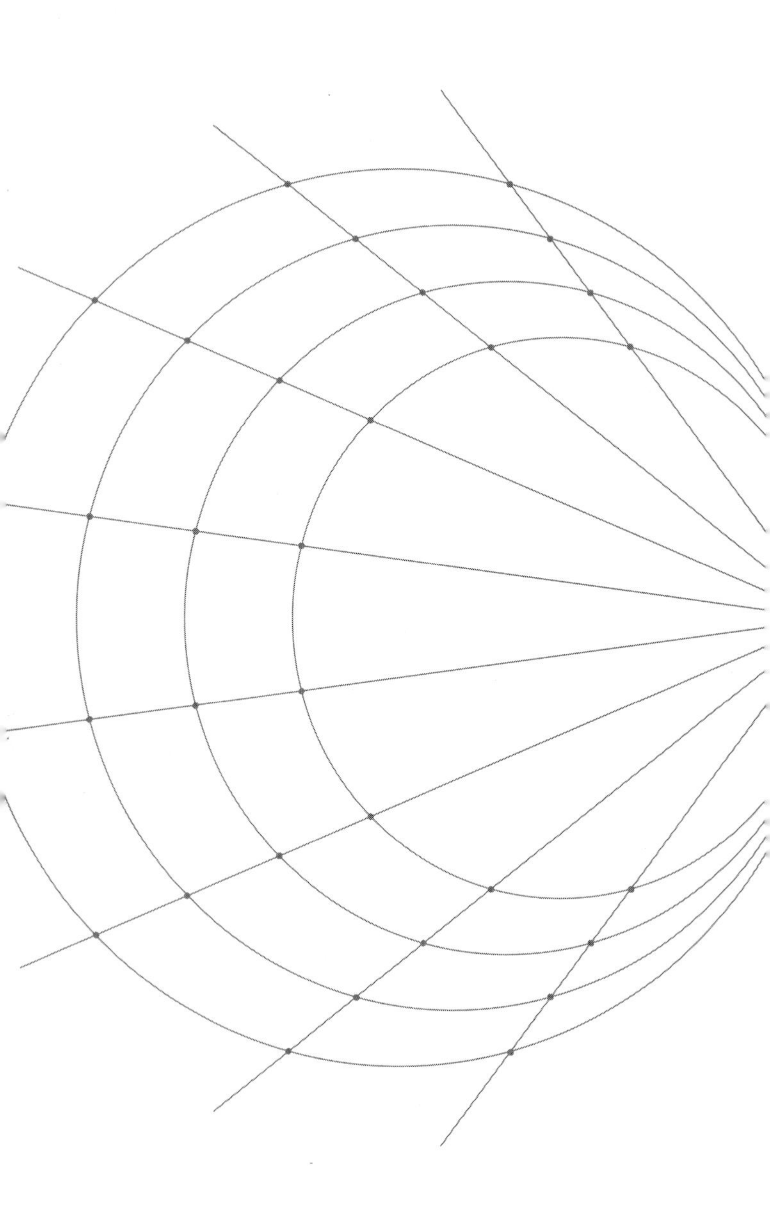

Cecil Balmond

Number 9

The Search for the Sigma Code

Nine Fixed Points in the Wind

Prestel
Munich · Berlin · London · New York

To Ruth and Hugh

Prestel Verlag
Königinstrasse 9
80539 Munich
Tel. +49 (0)89 24 29 08-300
Fax +49 (0)89 24 29 08-335

Prestel Publishing Ltd.
4 Bloomsbury Place
London WC1A 2QA
Tel. +44 (0)20 7323-5004
Fax +44 (0)20 7636-8004

Prestel Publishing
900 Broadway, Suite 603
New York, N.Y. 10003
Tel. +1 (212) 995-2720
Fax +1 (212) 995-2733

www.prestel.com

Library of Congress Control Number: 2008930233

British Library Cataloguing-in-Publication Data: a catalogue record for this book is available from the British Library. The Deutsche Bibliothek holds a record of this publication in the Deutsche Nationalbibliografie; detailed bibliographical data can be found under: http://dnb.ddb.de

Edited by Christopher Wynne
Cover-design by LIQUID, Augsburg
Book-design by Stefan Engelhardt, Mühldorf,
in corporation with Cecil Balmond
Printed and bound by PBtisk s r.o., Czech Republic

Printed on acid-free paper

ISBN 978-3-7913-4067-8

Preface to the New Edition

Twelve years ago a little boy entered my imagination as he hopped across the centuries and played with numbers. I began to see how the simple architecture of our decimal system could be constructed in secret ways – not a building project this time but an abstract one. On the surface of our arithmetic countless combinations of numbers take part in tedious and exacting calculations but underneath it all there is pattern, governed by a repeating code of integers. The Sigma Code reduces numbers to a single digit and the illusion of the many is seen to be but the reflection of a few. This is not a book on maths: this is a book for anyone who can carry out simple sums in their heads, and who won't be short-changed knowingly.

When *Number 9* first came out I received mail from many who played with numbers. They chased patterns; some had special numbers and even mystical systems. I was tempted to write about numerology but resisted. I wanted to write about the intricacy of what the numbers actually do and leave the reader to wonder about the larger irrational that seems to hover around such constructions.

If I were writing this book today the numbers would have featured in a wider context of structuring nature's patterns, and also playing the role of animator in algorithms that create unique architectural forms and shapes. I would also include my previous research into other base systems. But this book was a first step which came from a child-like urge, like playing with building blocks, to build out of our numbers – just the simple 1, 2, 3, up to number 9.

To begin the search . . .

It is like a detective story, gradually unfolding, because that is how it happened; not just for me but for Enjil, the boy mathematician who discovered the secret behind the mysterious world of numbers. In his case though he would have come to it quickly – but I laboured long many a night to find the answer. The thing to do is to follow the path until all the clues are in place and let your mind run free. It is only then that you find what the young master saw: the fixed points in the wind.

I had little to go on, just the stories without any detail, more legend than fact about how the numbers worked in secret and, in particular, the magic of one number.

My grandfather first told me about those stories and the special drawings and number shapes. Only the remotest people of the mountain village knew about these, he said, but it was all too distant and far away for me to give it any further thought. I was at university and more interested in working out how to put a rocket on the moon; playing with arithmetic was not for me – my mind was occupied with higher things.

Then one day I came across a children's book on numbers. One . . . two . . . three . . . . My eye went over the figures. Suddenly I saw something. There were hidden patterns; the old man's story about secret numbers came back to me and I became curious. I started to look into these simple ideas and the more I searched the more fascinated I became. Something was indeed going on underneath the surface of arithmetic and what appeared as a unique calculation to the outside world was something quite different when viewed from below. Looked at another way, six and six was not necessarily twelve but something much more exciting – the number 3, of a secret code.

I was won over. I began to look at numbers differently.

I used the special code I had stumbled on to find out more and track what Enjil had done; I spoke to people and read books; I looked for the answer to the riddle that made the Elders blanch and stir uncomfortably on that fateful day of the Examination. And I finally visited the mountain villages where the pupils in the schools stared at the pictures before they did any calculations, in order that they may inspire themselves for the rigours of the task ahead. The pictures of course had been added to and decorated heavily but at their centre stood the original spirit of Enjil's drawings, powerful and beautiful.

Through Enjil's investigation and my own research I have learnt many things I was never taught at school. When I think of the effort and the monotony I went through learning by rote, suffering numbers as necessary evils, I shudder. No teacher talked of the spirit of numbers, no teacher showed me the shape of a number. No one introduced me to a secret code that made lightning work of numbers and opened up worlds of wonderful possibilities hidden from the day-to-day grind; I found that other peoples too, in ancient times and in other lands, understood numbers as secret and special and alive, and not as mere counters, not just fodder for tiring calculations.

So I set out on the path Enjil took.

My own labours overlap his and our two stories have now become one. But that is how a personal search should be – with the spirit of that first discovery reaching out and embracing one until no difference can be found between one's own research and the inspiration that was first taken in. What was the author's becomes yours – which must be the meaning of original, something that embraces and absorbs all those inquisitive enough to enquire of others' inventions.

And it is in this spirit I dedicate the journey to you. Follow the clues, build up the jigsaw piece by piece and make your own investigations; become part of the search.

Go back in time and let the free spirit in you enter. Talk to it, play, ask the strangest questions.

Start to count again in the simplest of ways,

one, two, three, four … up to nine.

You need to do this, but you will also need nine clues.

And to begin with there is the story of Enjil himself, the talisman I conjure up whenever I think of numbers and of the fixed patterns that turn in the wind.

Nine Fixed Points in the Wind

Part One

Movements of Nine

Part Two

Nine Fixed Points in the Wind

Part One

Talisman

Enjil slept uncomfortably, his mind full of torment in fear of the Examination to come. He was standing in front of the Elders, those of the supreme rank, and he had nothing to say! He had not found a proof or a clever hypothesis to place before them on this auspicious day. And the day could not be put back – it marched right up to him, dragging him out brutally into the open, while the Elders, in their crimson robes, sat at the high table waiting for him. They motioned him to come up. He climbed up the steps and went to the blackboard and picked up the dry chalk in his wet, nervous hands.

Villagers crammed the square to see him perform. Word had gone out that the boy with the limp had magic powers; for when he lay dying from smallpox a strange bird had suddenly flown in and settled on his fevered brow, pecking at it. Superstition said it was the devil who seized a person's brain at such times, to give out great powers only to suck it back again at the moment of death, to prevent that tender soul from being re-born. And amazingly, as the bird flew away Enjil recovered and began talking in strange languages and writing down sheets of numbers, confusing everyone with ideas that they had never heard of.

Stories travelled through the mountain communities about the boy with the pockmarked face. How could the child be so precocious if not for some supernatural power? (No-one mentioned the devil's name for that would bring bad luck down on all of them!)

Enjil's beleaguered parents took him to the temple and gave him away to the priests, who in turn gave him away to the Academy and the Elders, these same Elders who now were laughing in his face – "Where is the cleverness that brought you here?", they mocked. "You insult us with your silence. We know who you are; The Devil's Child, a horrible trick; a wretch we now must throw out from this place of learning – Go! Only scholars are admitted here, true scholars like Vivek , who can work out the symmetry of magic squares better than you." The insults came thick and fast like poisoned darts thrown at him, tipped with venom.

But the boy had nothing to say, his hand stuck fixed in the air with the chalk wet in his fingers. And the villagers became angry at being deceived. Provoked by Vivek, Enjil's older rival for the title of Master, the villagers aroused themselves into a riot. As the mob hurtled towards him Enjil woke up with a scream, his heart racing. The sweat poured off him; he felt he was dying. But framed in his window the night moon shone brightly, the wind rustling the leaves like waves washing the shore. There was no mob, there were no accusing Elders. Everything was quiet and peaceful. There was really nothing wrong with the world, or him. The Examination was still days away; and he was well prepared!

His thesis was done – he had a proof written out in the higher algebra which his mentor had said would easily give him the title of Master; it was the sort of thing the Elders would like, for it was similar to the studies each of them had done. "It is not about being original, Enjil", his mentor said, "for you must not sit uncomfortably in your superiors' minds. If you present something they do not understand, or agree with as high learning, they will fail you. Conform, and then privately get on with your real discoveries. That is what we all do." His teacher shrugged at the way life was at the Academy and worry grew on the old man's face at the thought of what his stubborn student might do. But Enjil had listened, he had conformed – his thesis was as fine a piece of complicated mathematics as one could wish for, deliberately put together in an obscure way so that difficulties abounded in every line of the argument. In truth, Enjil had a much simpler proof; but it would appear too easy. So he had put it to one side and applied himself to obscurities in the demonstration of his thesis, knowing this approach would be more favoured by the Elders. And the peace of the night said he had nothing to worry about but go back to sleep and wait happily for that day of the Examination.

As he lay there looking up at the moon, his mind began to wander over the ideas he really loved to think about, like the expansion of π. Was there a pattern to it? Or how many unlike squares could fit into a rectangle or another square? This was a problem no one at the Academy could solve, though

Enjil had come near. Was every even integer the sum of two prime numbers as ten was the sum of three plus seven?

And there was more – the multiplication patterns along diagonals used by the Chinese or the elegant ratios used by the ancient geometers that gave beauty and shape to the spiral and to the growth of the leaves around a stalk or the petals in a flower.

As he roamed through the numbers in his mind a strange thing then happened. A moon beam suddenly reached out to him and dropped to Earth and turned into a shining woman. "So here you are!", she said smiling. "I have kept looking out for you. I find you here of all places, in a musty old Academy or is it a temple?" She wrinkled up her nose at the small room he slept in. The woman stroked his leg – "Does your leg hurt? I saw you as a child, limping, dragging your leg through the sand, making patterns that were wonderful – my friends still talk about you. I am Soma. Do you still play such games? You gave me and my companions a lot to think about that day, about the possible patterns in a matrix, instead of just the straight across and up and down. And now here you are, almost fully grown, yet still only a boy and sitting for the Examination of Master! Hah, that will sound fine – Master Enjil, Master Mathematician! How'd you like that?"

She patted him on the head and stroked his hair. Enjil could see the beautiful colours in her eyes. Her energy flowed into him. She took his thoughts away and she spoke to him:

"There are the cumbersome proofs you follow that only the few will ever understand. Your proofs are carrion for those vultures, the Elders. So why don't you do something else, something amazingly different for your Examination? You are capable of it! Why give the elders what they want, such a narrow outcome from your learning; why not something that everyone can enjoy?

Imagine even the villagers following your every sign on the blackboard, understanding it and seeing a simple but great truth unravelling right before their eyes. Something that lies under their noses, wouldn't that be fun?

How about drawing the many different shapes of squares it takes to fit in just one square. Ah – I know you were thinking of this already. You see I know your thoughts – so I count them out. And I'll spoil it for you anyway, I'm going to give you the answer; twenty-four! Yes, don't look so amazed, it takes that many different squares to fit into one square. You want to know the size of the square that allows this to happen? Ah, that is harder. That you must work out for yourself." She smiled teasingly and added, "Do you know the answer to the simpler problem; how many different squares does it take to fit in one rectangle? I'll tell you. It's nine! Nine unequal squares go to make up a rectangle."

"But why bother about geometry; why not something simpler than that? What is it that everyone knows and feels expert about? I'll tell you. It's numbers of course! Imagine the village folk clapping hands and cheering as they see you, their

new master, working with the humble materials of numbers, the tools they use every day of their lives to count coins and goats and sacks of grain from the harvest. Let your mind dwell on this: numbers! Take the most simple ones. Think of their make up. Don't be afraid of the Elders; they are not bad men, but men with too much oldness stuffed into their brains. Take them back to their childhood, let them smile again and hop, skip and jump through your constructions; what do you think of that? You smile? Do I take that as yes? Good! Then here is the riddle you must solve. And remember I will be watching, now that I have found you; but I won't help. I will only give you a signal when you succeed. Remember the only thing that will slow you down or stop you is the amount your mind has grown up to be like the Elders, the brain of an expert. The problem I set is for a child, with a mind that in innocence questions everything and finds new beginnings. I too, believe it or not, am like that. I live in the moon and each night I set a new day. The turning of a fresh beginning uplifts me. It keeps me from falling into endings and dull repeat reasonings. Think about that. In your dreams I will speak to you, I will help. My spirit will be with you. But now I must go for I have to set another day. And this is the question I leave you with:

What is the fixed point in the wind?"

The woman withdrew along the moonbeam and vanished as if she had never been.

Enjil sat up. He looked hard at the moon, staring into the white disc of cold light – and the moment of magic vanished. Was it a trick? Was Vivek his arch rival for the title of Master trying to hypnotise him from a distance? If he listened to the woman it would be like suicide. Standing in that great open courtyard and speaking about simple things that did not need proofs would have the Elders laughing at him, baying like jackals at his feeble efforts. Certainly he would be thrown out. This must indeed be a hex put on him by his enemies!

But what was the fixed point in the wind? The question intrigued and teased him. How can something be unmoving in the swirling wind; what was its fix, if indeed there was such a point? For hours he lay awake struggling with these thoughts until his tired brain came to a stop and wanted rest. Finally Enjil fell asleep. The woman in the moon entered his dreams.

And he woke up with a new conviction – the doubts and torments of the night well behind him. He whistled and even smiled at Vivek, winking at his arch rival as if to say he had prepared a brilliant proof. Enjil laughed to himself when he thought of the title he would introduce to his Elders on the day of the Examination. He would wear the yellow robe of scholarship, go up to the blackboard and announce in his most stern voice the customary words, "My respectful Elders and Seniors. I submit for your Examination and proper adjudication this thesis I have now prepared for the award of the Most Expert Master of Mathematics, the honour I now seek, and pronounce as the title of my learned subject: 'The Fixed Points in the Wind'." He could see them writhing in agony, splitting their sides with laughter and thirsting for his blood. The images made him break out in a cold sweat. But a small voice spoke inside saying, "Don't be afraid – of course you can do it – after all it is an easy question – so solve it like a child, think like one, just like I said."

To solve the riddle Enjil went to a secluded spot and sat in the shade of a banyon tree and blanked everything he knew out of his mind. The great blackness descended. Nothing moved. Shadows went into deeper shadows, layer into layer. A black disc grew. First as a dot, then a circle, then a rushing blind movement. Then the numbers came out, tumbling one over the other, rolling the patterns over in his head. There were the star patterns, zigzags, squares, cubes, seesaws and the weaving patterns going in and out, all twisting over each other. Mindful of the woman spirit, he looked at the simplest numbers; he followed their trails, the white and black patterns, some, dotted with colour, moving like the wind, changing shape and turning all the time. And there in the simplest patterns were points that did not move or change, no matter what the numbers were. And they were fixed points. When the woman in the moon had talked of the wind Enjil knew she must have spoken of numbers, jumping over each other in gusts of multiplications or blowing steadily in ordered breezes of additions and subtractions. But within these patterns, there was one number and point through which all the others seemed to gather – it indeed must be the fixed point. Could that be the answer to the riddle?

Then Enjil opened his eyes and composed his thesis. It was so simple that he laughed out loud. Never mind the Elders – they would have to love it, for he would draw the movements of all numbers in one simple diagram. What was clever about it was the method he had in his mind. He would take the secret code the Elders knew about but had never thought of using to look beyond their rituals of prophecy, for they would take the letters of someone's name and use such secret numbers to divine the character of that person. But Enjil vowed to go beyond this.

That night he wrote out his thesis. When he had finished he went out into the deserted courtyard and held up each page to the moon. "Look", he said to the woman in the moon, "I have finished – the task is done. I pray these are the answers."

A shape seemed to move across the face of the yellow disc though he could not be sure. But the woman did not appear. The wind picked up, the night chill made him shiver. Tomorrow was the day, and Vivek would be hoping for his downfall; and his teacher would fret to the last moment as the Elders assembled, sharpening their wits in readiness to humble the nervous candidate. The crowd would gather and settle. Everyone would be waiting, watching.

Still, Enjil was at peace. His ideas were simple, innocent; he believed they would shine through no matter what. While Vivek performed great feats of algebra Enjil would offer to the Elders and the gathered crowd the four precious mirrors of arithmetic. They could all laugh or puzzle at the reflections he would show them, but in time their doubts would vanish or be blown away, just as the gusting wind cleans out the dirt lying on the ground.

Suddenly a bright light flared, lighting up the compound for an instant, and then faded. Had the woman come to him and acknowledged his answer? Enjil rolled up the pages of his thesis and held up his hand, just in case, to the night sky in salute and farewell and then went to bed. He slept the peace of the innocent, a fixed point himself in that night of swirling anxieties and jealousies. Tomorrow would be a new beginning.

I wear Enjil's Talisman now, whenever I calculate and look into numbers, and remember the mathematician who was a boy. He inspires me to look afresh at things.

Enjil went on to be famous. He surprised the Elders with his arguments of the fixed point in numbers; he astonished the crowd. And if not for them cheering as they did at what was being drawn on the board, the Elders would surely have failed him – the old men being insulted that there was no high algebra or long-winded, obscure complications to be resolved. Enjil's workings were just basic arithmetic, they grumbled, which even a nine year old could follow, they whispered to each other. It was laughable; it was ridiculous and all too simple. But the crowd cheered and cried out, "Master, Master," so many times, that the Elders gave in. On the boy's shoulders they placed the purple sash of Master embroidered with winding circular motifs in gold thread, and then they held up Enjil's hand to the crowd who in turn roared, "Mas-Ter, Mas-Ter". The four syllable chant rocked the compound. When the dust cleared and the noise subsided and the courtyard was empty, the temple still hummed long into the night with stories of the new Master's cheek and sheer luck.

So Enjil went on to be the real Master of the Academy, outshining everyone and everything. He proved many things, much of it beyond the best minds in that Academy of high learning. But he kept faithful, from what we know of his teachings, to the simple and straightfoward, and always the beautiful and intriguing. The Master saw patterns where others only saw calculations.

Then disaster struck. In the wars that ravaged the country the Academy was set on fire and the intellectuals speared to death. Enjil and his papers and all the great library of learning in the Academy were lost forever. No one found the young mathematician's body. Soon the story went out confirming Enjil as a spirit child, one that visited Earth now and then to remind us of the greater glories that hid elsewhere. Others said it was the work of the devil, who flew in to collect the soul that he had claimed for himself long ago, when the little boy had lain dying from a fierce attack of smallpox.

Whatever the truth of it the crooked smile and limping walk of the Master was no more, severely missed by those who loved him. Those who were there on that day of the Examination told others. And the words and diagrams spread. The ideas travelled from community to community; changes and additions were made along the way. But the basic structure Enjil proposed is still there in the teaching.

Modern ways have swept across the culture of this great land and hand calculators and computers have taken away the simple romance of numbers, but in remote parts of the highlands, in locked away villages, young mathematicians still look at Enjil's patterns and meditate for half an hour before doing any serious mental computation.

What follows is a trace of the young Master's working, gleaned from private study and old village stories. With Enjil we move towards finding the magic of numbers and that special point, which though full of movement itself, remains unmoving and stationary; just as a fixed point does in the wind.

The Four Corners of the Earth

Clue 1: Corners of the Earth

The first clue is in the wind itself; where does it come from; where does it go? During the summer the land mass heats up and the cool sea wind is dragged in from the South-West and with it the monsoon rain. In the winter, the cold ice of the mountains chills the land but the Northern seas are warmer; the wind blows out in that direction, towards the North-East.

In other countries similar changes take place between the seasons. There is no fixed pattern. The wind moves in all directions, coming or going, its movement held by the swing of summer or winter. The season's alternation gives the dynamic. Local features intervene affecting fast or slow, gust or breeze, and in small pockets different directions take hold. But we can give a sense of order to this swirling change by mapping eight directions.

First we mark the cardinal points North-South and East-West.

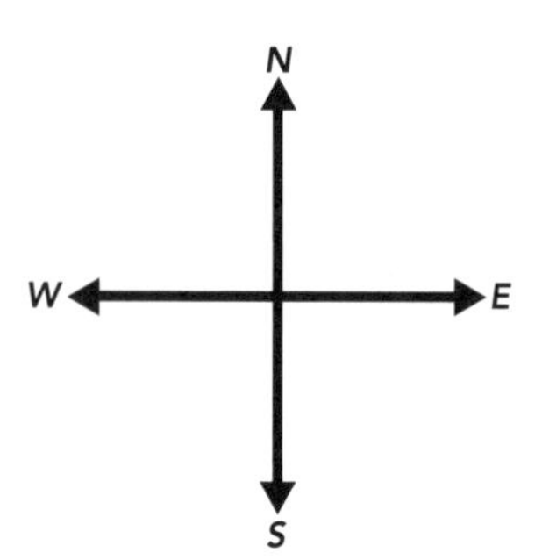

Then the directions of North-East, North-West and South-East, South-West, are marked as four more corners.

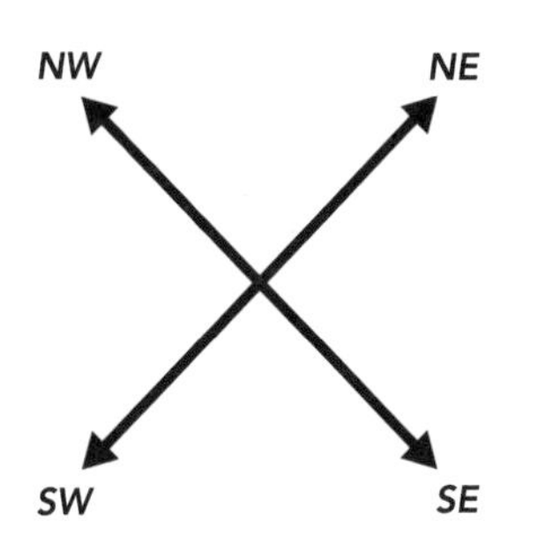

If the points of the arrows are plotted on the ground there are now eight points that corner the wind.

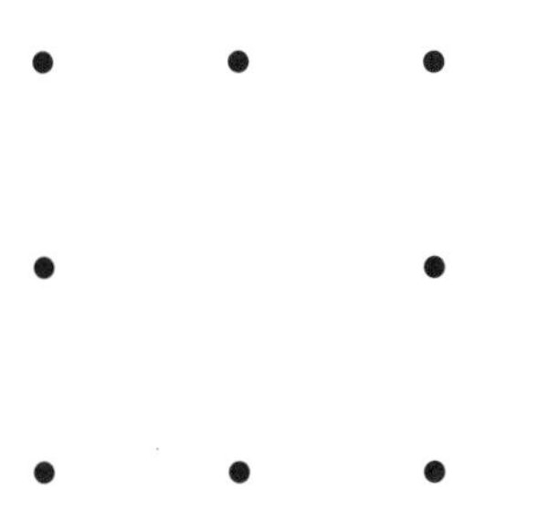

But these points only have meaning if there is a centre, that is, the origin of the framework. We need a ninth point, the centre of a star that radiates in eight ways.

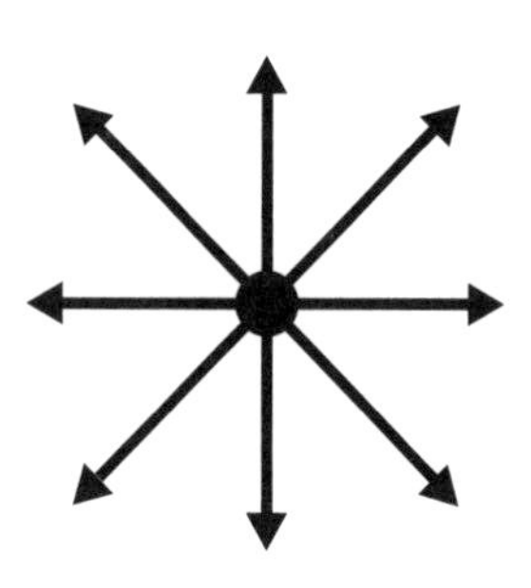

Whilst eight points give outer meaning, the ninth point in the centre of organisation makes a strategy possible for mapping direction.

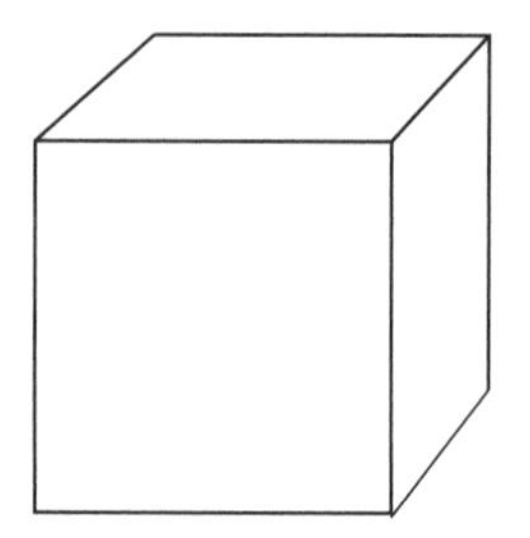

A long time ago the cube was the symbol for the world. Square in elevation and plan, of even character, the cube was the essence of the solid body of the earth and stood as a symbol for its substance.

Eight corners of the cube gives sufficient outward consolidation and mapping to shape, but hidden in the construction and implicit is a ninth point, the fixed point through which all else happens. In this defining point of the cube, where all its mass is supposed to act, a centre of balance and symmetry occurs at the secret heart of volume.

Is nine special?

At first I did not think so; it was not obviously a number I thought of as auspicious. But then I found that Buddhist relics in stupas are buried at a central ninth point, around which eight small buddhas sit marking the directions of the world.

I found that the number nine seems to be a point of initiation and departure, a beginning and an end.

In human life it is nine months from the moment of conception to the birth of a baby. And at a practical level there are nine passageways in and out of the body.

There are nine parts to heaven and nine planets!

I found stories and myths full of the number nine. Yes, there are other numbers that are used in old legends; numbers like 7 are part of the Bible stories and its lucky mark holds in many myths of good omen. And looking through the reference books on myths, number 3 comes over as a very powerful number because the first organisation of 'things' seem to be in threes: the Three Bears, the Three Blind Mice, the Three Kings, and the Holy Trinity. But 3 of course is the prime mover of 9. It is the thrice magnification of three that makes up the power of nine!

So the references grow as one probes; number nine seems to have an uncommon hold in the early stories we first made to describe the world. But before we enter this frame of archetype, we must first look briefly at the characters of the numbers themselves, their symbols and their qualities and witness their hold over our imagination.

Numbers

In the world of arithmetic numbers are just counters, they are markers in calculations, having no other property but to represent their own value. But if I look for relationships between the numbers then something else happens, the numbers come alive.

The great discovery of Pythagoras that 3, 4 and 5 were not just plain numbers in a sequence but that

$$3^2 + 4^2 = 5^2$$

defined right angledness, is an exciting idea to contemplate. But when I find that

$$1^2 + 2^2 = 5$$

a whole new layer of intrigue and speculation grows over the previous answer. The question is asked: are there other such patterns in the hypotenuse of right-angled triangles? When the answer is yes, we are led into a world of open-eyed wonder that such relationships even exist. We enter a world of number theory and find strange families and fascinating behaviour patterns. The search is a pure reward because we find out patterns for their own sake; the numbers seem to grow in character and develop their own sense of mystery.

And if we contemplate what a number means to us, in our imagination, a further layer grows as to what its colour or shape is. Is it awkward or comfortable, powerful and royal or cadaveric and evil? If we just close our eyes and think of numbers, we open all of our senses to them. I can touch them and 'feel' their texture or inward force. Are they magical?

Uniquely, they touch us in different ways, exactly according to our own make-up. Some are steady and robust, some vibrant and exciting, others dynamic and lucky.

What is intriguing is that the smaller a number is the more potential it seems to have. Large numbers do not seem to be as interesting as small numbers. Smallness seems to focus a number better in our mind's eye. And the smallest we can make them are the single digits, one to nine.

1 2 3 4 5 6 7 8 9

These numbers have fantastic potential, for from here, in their purest form, they go on to structure the gigantic surfaces of all the combinations of numbers, and Enjil's discoveries lead us to find surprising shared characteristics even between these numbers.

But I am now getting ahead of myself; that story has still to unfold. The clues must be played out and chased, and puzzled over.

First, let me introduce the characters themselves.

Odd and Prime

3

THREE is special – we are 'three times lucky'. Most important categories come in threes, like Mother, Father and Child, or the Trinity of the Christians: Father, Son and Holy Ghost. The atom is classified in three main parts as Proton, Neutron and Electron, and all matter is either solid, liquid or gas. Three is also the mark of the eternal triangle, the first definition of a two dimensional area according to books on geometry.

5

To Sorcerers and Magicians number FIVE is the most powerful – five is the mark of the pentacle, a five pointed star drawn by extending the sides of a Pentagon. Five surely is in the possession of the occult. And the Pentagon is the geometric figure in which the golden ratio of classical art and architecture is found most.

7

To others SEVEN is the lucky number, associated with mystery and mysticism. In the world of the Buddhists and the Taoists seven spirits put the world in order, the seventh putting the pillars of the Heavens in place to complete the act of creation sublime.

3, 5 and 7 are all odd numbers – they are also the first prime numbers. And luck and power are associated with these numbers.

Number ONE is the beginning of all numbers. Every number is an accumulation of units of ones. Since one is the source, it is the binding principle of all numbers and in a sense is beyond counting. One stands for identity. We stand upright in the individual figure of one, and we stand together as one.

Even Numbers

Like one, the number TWO is also important as a basic principle; it provides all doubling and evenness. It smoothes out all angularities and oddness, for twice an odd number gives an even number. If one is unity, then two is duality and provides a balance between opposing halves. Two is also unique amongst prime numbers – it is the only even one.

To the ancient Greeks FOUR was the symbol of the Earth. Standing 'four square' was also the measure of Pythagorean justice. Christians mark a figure of four points on their person in the sign of the Cross. A curious property of 4 is that any odd number squared, when divided by four, leaves a remainder of one.

6

For centuries mathematicians praised the number SIX as perfect – that is a number whose divisors sum up to the number itself. (Six can be divided by one, two, and three and these numbers when added come to six.) Perfect numbers are rare and very hard to find. Six is the number of days it took God to make the world. Six is the number of sides of a hexagon, also the first shape of geometry found in nature, as in the honeycomb.

8

To the early Greeks a person was a full chord of eight notes. And number EIGHT to numerologists is an auspicious number having affinities with twos and fours. There is also a curious relationship to number four in the division of numbers; all prime numbers when squared and divided by eight leave a remainder of one!

So we skip through the basic digits very quickly and have a thumbnail sketch for

1 2 3 4 5 6 7 8

But where is number NINE?

9

Seen as last in the line, before zero slips in and we start counting again, is nine a last stop charlie? Yet its stepping stones (3) and (6) seem to have a special power.

Three and six blaze with luck and perfection. And nine is also three times three, magnifying the power of three. Perhaps after all nine may come to hold a special promise.

3 + 6 = 9	*luck + perfect = ?*
3 × 3 = 9	*luck × luck = ?*

Nine is also twice four plus one. But four is equal to two plus two so

2 × (2 + 2) + 1 = 9

Nine seems to be a number in qualitative terms full of inner fortifications: luck, perfection and doublings. Perhaps we glimpse some of this in the arrow tips of the wind and in the binding symmetry of a cube, both built on the organisation of an unseen but vital ninth point, the central point. Maybe nine is the core between all the corners of the world? And what takes us to the corners of the world is the stories. Once upon a time we gave shape to these stories with numbers; in the traditions that grew, the numbers became archetypes, with a symbolic power all of their own.

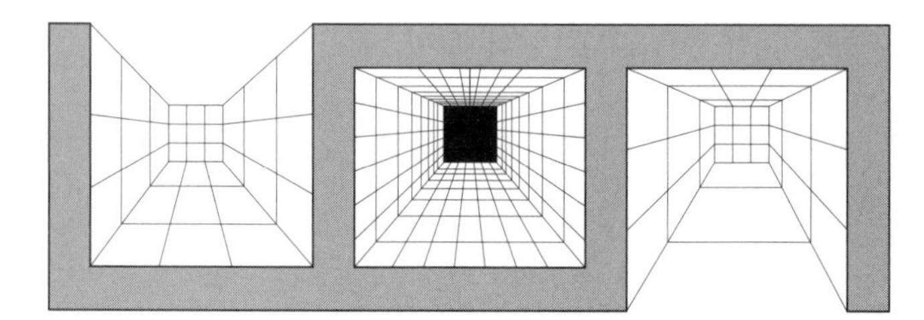

Archetypes

Clue 2: Archetypes

First there is the archetype of symbol.

A thousand years ago the number nine was fixed by Arab mathematicians in the notation we now use, the curling in of a line that turns continuously.

But Arabic mathematicians inherited a Hindu script in AD 876 and an inscription from Gwalior which showed the number 9 like this:

The Gwalior inscription in turn came from a development of the earlier Brahmin numeral for nine, from the second century AD. And before that came the older histories and markings of the Romans, Hebrews, Greeks, Chinese, Babylonians and Egyptians.

Figures of Nine

Egyptian	
Babylonian	
Chinese	 九
Greek	 θ
Hebrew	 ט
Roman	 IX
Indian	 ९
Arabic	 9

Then there are the tales of great spirits and strange forms that govern the beginning and endings of things. They are the stories across cultures which invoke nine as a magic ingredient in rites of passage.

The Seeds of Nine

Yang formed the Heavens and Yin formed the Earth. And Pan Ku, who was in the middle, changed his form nine times a day, sometimes into a God in heaven or into a saint here on earth.

A dead person must cross the Chinvat or 'grading' bridge, which is as wide as nine spears laid end to end for the just, and as narrow as the finest edged razor blade for the wicked.
(From Persia)

The dwelling place of the dead is not easy to get to; there are steep mountains to climb, terrible deserts to cross and poisonous snakes to confront. The wind pierces the body and the weary soul looks for a final resting place but first must cross Hell's frightening rivers, all nine of them.
(From Mesoamerica)

Homer wrote in the Odyssey:
At the age of nine, they were nine orbits wide and nine fathoms tall and they threatened the Immortals by bringing the tumult of war to Olympus. They wanted to pile Mount Ossa onto Olympus and Mount Pelion with the restless leaves onto Mount Ossa in order to mount their assault on Heaven.

Haephaestus was a magician, unique in the world of the Gods, but he was lame and his mother thought him ugly and dropped him from Olympus. He was brought up for nine years in an underwater cave by Tethys and Eurynome and he there learned handicrafts and the art of jewellery and the making of necklaces.
(Iliad XVIII)

It took Vulcan nine days to reach the island of Lemnos when he was banished from Olympus.

The voyages of Odysseus lasted for nine years before he arrived home.

The duration of the siege of the City of Troy was nine years.

The Ark of Delucion was tossed about for nine days when it became stranded on top of Mount Parnassus.

When the fallen angels were cast out of heaven for "nine days they fell." (Milton, Paradise Lost)

Niobe, the wife of Amphion, King of Thebes, boasted of the number of children she had. Latona raging with mad jealousy called her own two children to seek a terrible revenge and Niobe's loved ones were destroyed. For nine long days Niobe lay beside her children, weeping for them, before their bodies were buried into the darkness and she was turned into stone.

In the old ways of the Norse, Odin was a great traveller, and he wanted to understand everything. But wisdom could not be bought by gold or silver, so he gave his eye to the woman who guarded the fountain of Mimir, so that he could truly see. And Odin discovered the runes, the sacred writing which allowed thought itself to be set down and passed on. To do this he hung from the tree which was battered by the winds for nine long terrible nights, pierced by a spear. Without having eaten or drunk he picked up the runes shouting out he knew how to increase and prosper. He engraved the runes into wood carvings.

Heimdallr, the God of the Norse people, could see everything and never closed his eyes. He could hear everything, the grass climbing out of the earth and the wool growing on the back of sheep. He guarded the foot of the rainbow which led to the Gods. Heimdallr, the special one, was born of nine mothers.

There were nine Muses; they were the children of Zeus and Mnemosyne. They were not only divine singers but they were patrons of all intellectual activities of the times, including the highest, which was everything that freed man and gave access to the eternal truths. These included Eloquence, Persuasion, Wisdom, Knowledge, Mathematics, Astronomy as well as Poetry, Music and Dancing.

There were also nine virgin priestesses of the ancient oracle.

Dryden wrote in The Flower and The Leaf:
Nine worthies were they called, of different rites,
Three Jews, three pagans, and three Christian Knights.

The Jews were Joshua, David, and Judas Maccabaeus;
The Gentiles were Hector, Alexander,
and Julius Caesar;
The Christians were Arthur, Charlemagne,
and Godfrey of Bouillon.

Sometimes the nine worthies were referred to as being three from the Bible, three from the Classics and three from romance.

But nine covers Hell and the dark side in thrice threefold ways as well.

At last appear
Hell-bounds, high reaching to the horrid roof,
And thrice threefold the gates; three folds were brass,
Three iron, three of adamantine rock,
Impenetrable, impaled with circling fire
Yet unconsumed.
(Milton, Paradise Lost)

To make a charm the witches chant:
Thrice to thine,
and thrice to mine,
and thrice again to make up nine.
(Shakespeare, Macbeth 1 [iii])

Buddha was the ninth incarnation of Vishnu.

Vishnu created the world in three strides. He pushed apart the universe and placed the sky, the heavens and the earth in their rightful place.

Three times three, the trinity of trinities, gains select status then as the doubling and resourcing of special power.

$3 \times 3 = 9$

From ancient times number nine was seen as a full complement; it was the cup of a special promise that brimmed over.

The organisation of heaven:

Seraphims	Cherubims	Thrones
Dominions	Powers	Virtues
Principalities	Archangels	Angels

To the Greeks a person was a full chord of eight notes; the ninth was the all embracing sound of the deity.

So nine enters religion.

Allah is blessed with 99 names and the Feast of Ramadan is on the 9th month of the lunar year.

For the Cabbalists there were ten emanations of God; from Malkuth, the Kingdom Earth, a soul has to leave its earthly body and move upwards towards the nine spheres above to reach Kether, the creative Godhead itself. It is easy to see how early numerology assigned nine as the Royal Number, the mark given to high creativity.

In the Catholic Church there are nine First Fridays. In a vision from God St Marguerite Marie Alacoque, a French nun, was given the message that special devotion should be made to the Sacred Heart. On the first Friday of every month for nine consecutive months, those who attend Mass and receive the Sacrament of the Eucharist, pray that special grace is granted.

Novenas, which are said for special intentions, are for a period of nine consecutive days.

On the Sermon on the Mount there are nine categories called Blessed:
Blessed are the poor in spirit; the patient;
and those who mourn;

Blessed are those who hunger and thirst for holiness;
Blessed are the merciful;
Blessed are the clean of heart;

Blessed are the peace-makers; those who suffer persecution in the cause of right; and those who are reviled and have all manner of evil spoken against them in His name.
(Matthew 5, 3:12)

Jesus was crucified at the Third hour.
At the Sixth hour the world was plunged into darkness.
At the Ninth hour Jesus yielded up his spirit and died.
(Mark 15, 25:39)

Beyond the stories there are the numbers themselves:

the Great Year of Babylon;
the verses in the Rg-Veda;
the choirs of angels in the Book of Revelation;
the years in Hell for a Buddhist;
the gates to Valhalla;
and all the names of Allah.

The Babylonian Great Year is **432 000** years long.

Written thousands of years ago, that most sacred Hymn and first recorded poem of existence, the Rg-Veda of the first Hindus, has stanzas **10 800** in number.

In the Book of Revelation a choir of **144 000** angels redeemed the Earth.

For the devout Buddhist, Hell can last **576** million years!

Valhalla is said to have up to **5 400** gates to Hell.

Allah has **99** names.

When I look at these great and awesome numbers there is one surprise, they all have a secret; in each case the sum of their digits adds up to nine!

432 000	→	*4 + 3 + 2 + 0 + 0 + 0*	= **9**
10 800	→	*1 + 0 + 8 + 0 + 0*	= **9**
144 000	→	*1 + 4 + 4 + 0 + 0 + 0*	= **9**
576	→	*5 + 7 + 6 = 18; 1 + 8*	= **9**
5 400	→	*5 + 4 + 0 + 0*	= **9**
99	→	*9 + 9 = 18; 1 + 8*	= **9**

Deep down in the fabric of the numbers that describe and classify these great events is number nine, planted like a hidden seed. Throughout sacred literature this number keeps cropping up as the mark of auspicious events, acting as a trigger point of initiation or departure.

Even in the archetypes of our hands, nine is a catalyst to counting. Each finger can be divided vertically into 9 points, a series of 3 on each side and one up the middle. The points on one finger can then be allocated numbers 1–9. The next finger can be marked from 10–90 another 100–900 and so on, until with one hand, as the early Chinese did, one can count up to ninety-nine thousand, nine hundred and ninety-nine.

9 9 9 9 9

COUNTING FROM 1 - 99,999.

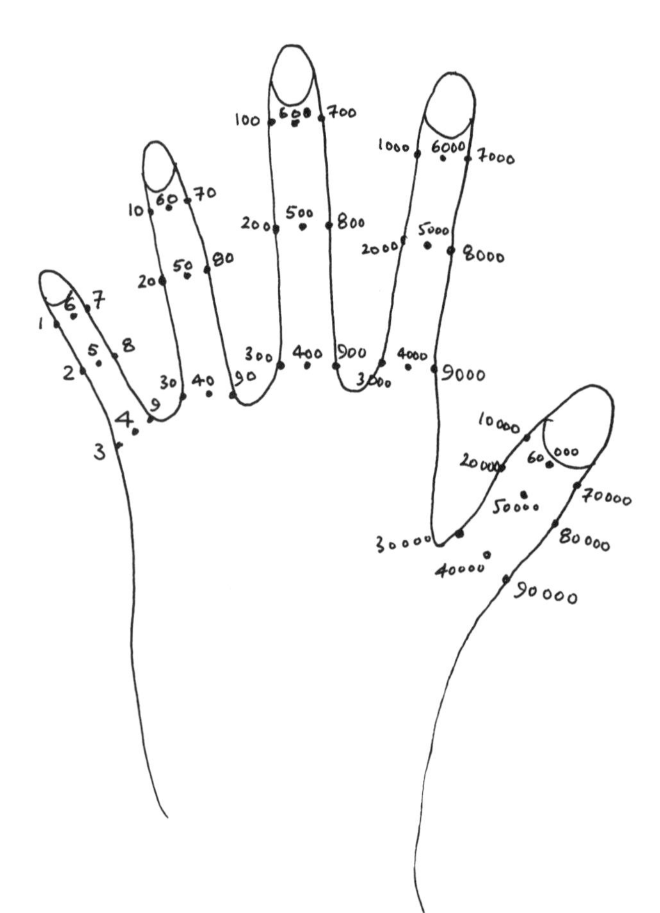

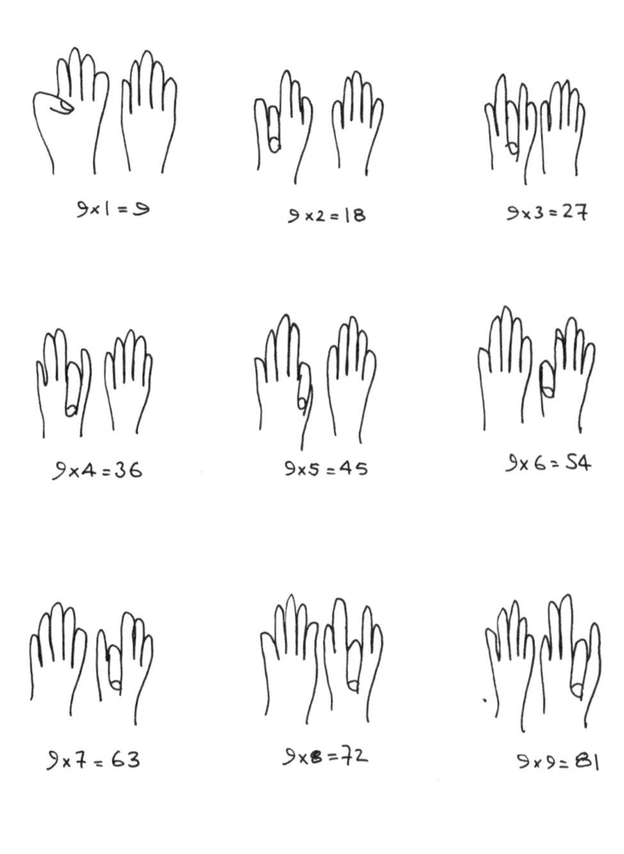

MULTIPLYING BY NINE

The First Magic Square

The Emperor Fuh-hi was sailing down the Yellow River and saw a large turtle in the water, on its back were strange markings. The marks made up the sign of the lo-shu, or first magic square, each dot standing for a number.

Today we arrange the lo-shu into the nine parts of a square and instead of dots use numbers to mark down the magic. The square is supposed to bring luck.

Whatever the direction of addition, vertical, horizontal or diagonal, the numbers in the square add up to 15. The square is $3 \times 3 = 9$ parts.

And the magic number 15 adds up to $1 + 5 = 6$

If you multiply the numbers across each line $4 \times 9 \times 2$; $3 \times 5 \times 7$; and $8 \times 1 \times 6$ and add them up, the total is 225.

Similarly if you multiply the vertical lines $4 \times 3 \times 8$; $9 \times 5 \times 1$; and $2 \times 7 \times 6$ and add them up the total is also 225. And $2 + 2 + 5 = 9$.

We have the stepping stones of nine ③–⑥–⑨ emerging in an intriguing way.

By travelling through the archetypes of ancient times we see the number nine as a mark of passage, of initiation and departure, and a number full of mystery. But these stories are all qualitative, and intriguing as they are, a much greater curiosity lies in finding out just what nine does in the world of numbers.

To find the real character of nine we have to enter a special world and peer into the four precious mirrors of arithmetic. There we must catch reflections and pursue the elusive nature of what turns out to be a chameleon amongst numbers, now you see it, now you don't.

But first, we have to play a trick with reversals.

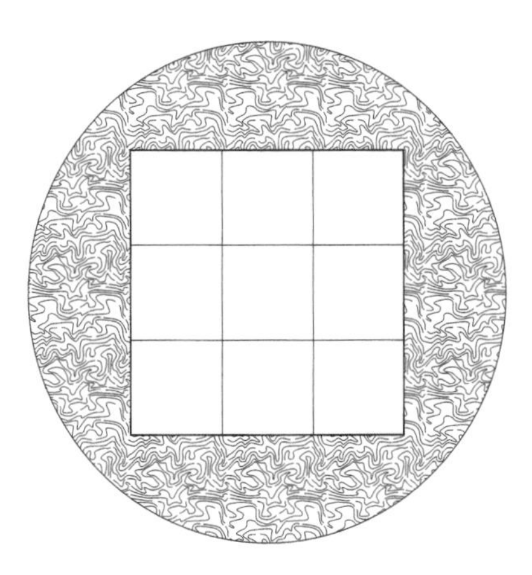

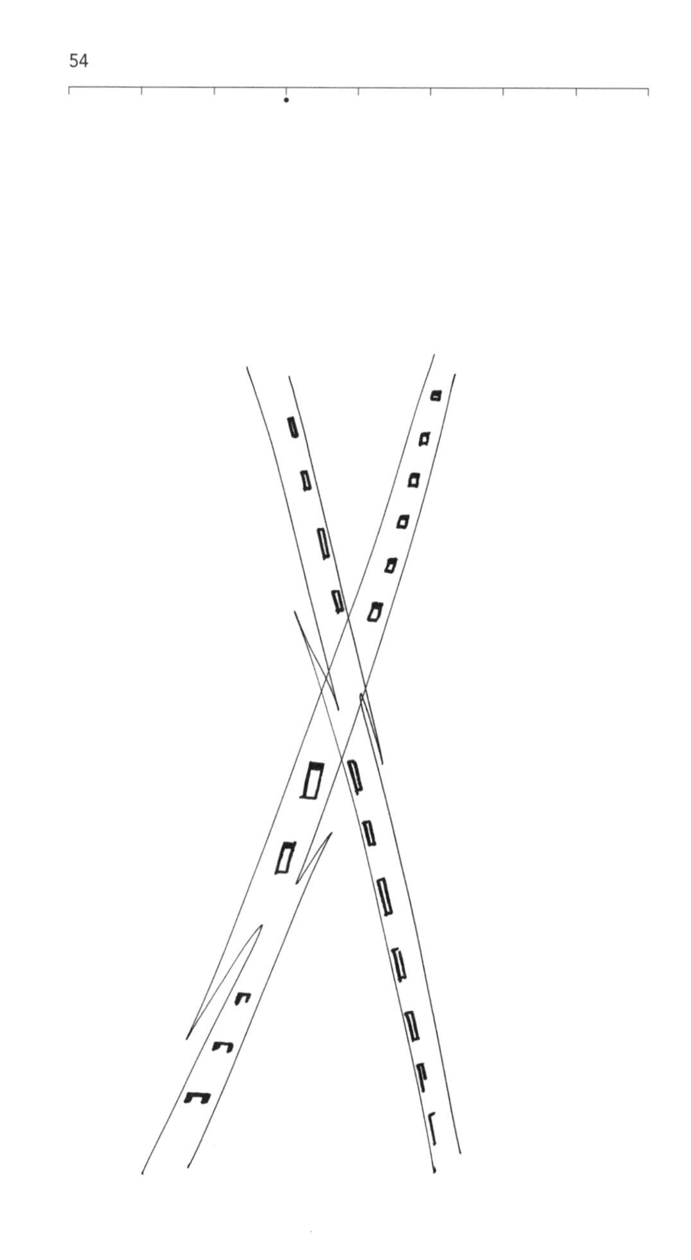

Reversals

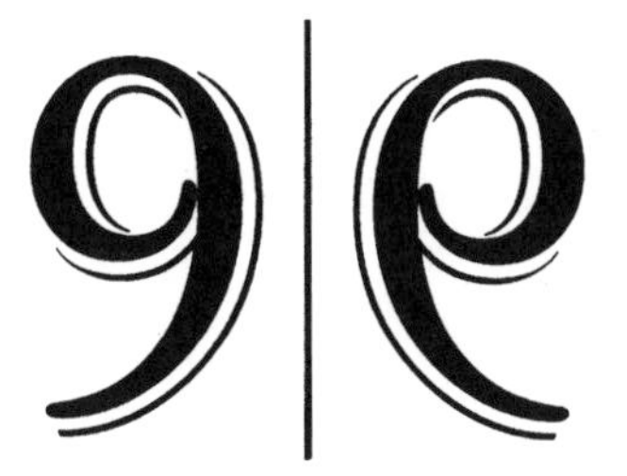

Clue 3: Reversal of Nine

A trick with numbers that the conjurers played went like this:

Take any three digit number, say 781.
Reverse it and subtract it from the original number:

$$\begin{array}{r} 781 \\ -\ 187 \\ \hline 594 \end{array}$$

Now reverse this answer 594 to 495
and add it back to the last sum:

$$\begin{array}{r} 594 \\ +\ 495 \\ \hline 1089 \end{array}$$

Whatever three figure number is used the number 1089 always comes up. The only rules to follow are that the first digit of the chosen number must be greater than the last digit, so that subtraction can take place, and that the 0 is not omitted if it arises after the first subtraction.

Examples

652	562	221
− 256	− 265	− 122
396	297	099
396	297	099
+ 693	+ 792	+ 990
1089	1089	1089

The answer is always 1089. Try it!

If the same idea is now repeated with any two figure number, always keeping the first digit higher than the last one we get:

Original number	81	41
Reverse and subtract	− 18	− 14
	63	27
Reverse answer and add	+ 36	+ 72
	99	99

Nine is somehow at the heart of the matter.

Where does 1089 come from?

Adding up the digits of 1089 gives:

$$1+0+8+9 = 18$$

and

$$1+8 = 9$$

Two 99s go up to make 1089 as follows:

$$\begin{array}{r} 99 \\ +\ 99 \\ \hline 1089 \end{array}$$

When 1089 is divided by 9 the answer is

121

a symmetrical number. Reading from left to right it is the same, as if a mirror runs through number two. There are further symmetries within this number because it is also the self replication or square of yet another symmetrical number.

$$121 = 11 \times 11$$

Number 9 is connected somehow with reversal but, at the same time, there is a hint of symmetry in the way 1089 is structured.

Task for the reader:

Try reversing four or five or six figure numbers in the same manner – see what happens with 4 321 or 79 685 or 528 613? Try decimals. Add the digits in the answers you get. Are they a multiple of nine?

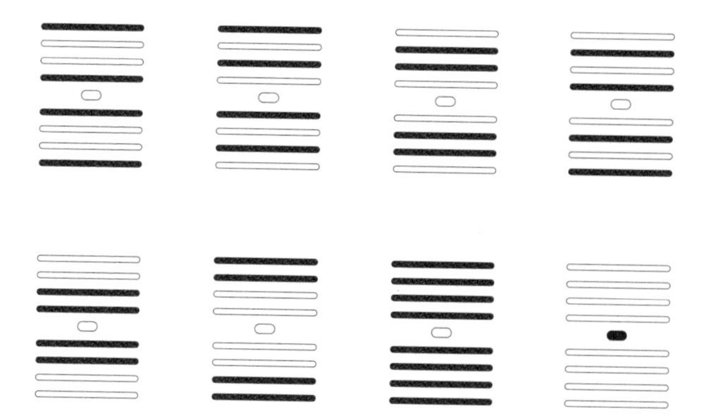

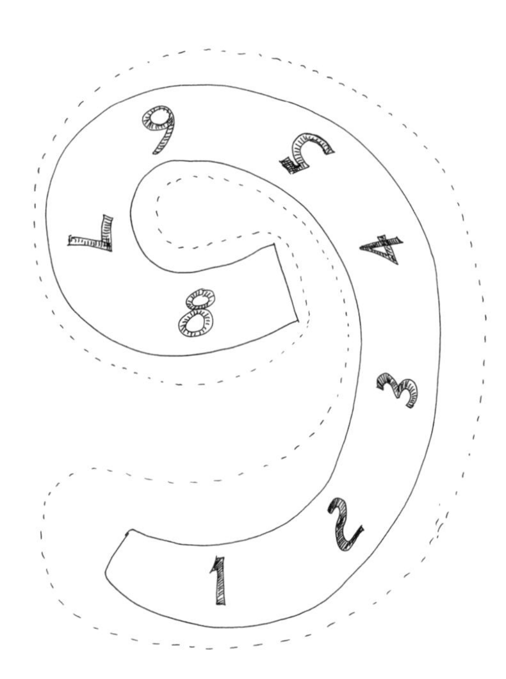
1
2
3
4
5
6
7
8
9

Secret Numbers and the Sigma Code

Clue 4: Secret Numbers

In an idle moment one day, as the sun shot its ninety nine thousand rays over my desk, I wrote down the integer values 1 to 99 in serial fashion, in rows of nine:

1	*2*	*3*	*4*	*5*	*6*	*7*	*8*	*9*
10	*11*	*12*	*13*	*14*	*15*	*16*	*17*	*18*
19	*20*	*21*	*22*	*23*	*24*	*25*	*26*	*27*
28	*29*	*30*	*31*	*32*	*33*	*34*	*35*	*36*
37	*38*	*39*	*40*	*41*	*42*	*43*	*44*	*45*
46	*47*	*48*	*49*	*50*	*51*	*52*	*53*	*54*
55	*56*	*57*	*58*	*59*	*60*	*61*	*62*	*63*
64	*65*	*66*	*67*	*68*	*69*	*70*	*71*	*72*
73	*74*	*75*	*76*	*77*	*78*	*79*	*80*	*81*
82	*83*	*84*	*85*	*86*	*87*	*88*	*89*	*90*
91	*92*	*93*	*94*	*95*	*96*	*97*	*98*	*99*

As I scanned the numbers I realised suddenly that the entire table was but a repeating pattern.

For example, the third and fourth rows in the table

19 20 21 22 23 24 25 26 27
28 29 30 31 32 33 34 35 36

reduced to the same nine values as the first row in the table when the digits in each number were added up. (If on addition the answer was greater than ten, then I added up the numbers again to get a single value. For example, 19 = 1 + 9 = 10 and 1 + 0 = 1; 20 = 2 + 0 = 2, and so on).

The table I had written out from 1 to 99, then reduced to nine rows of numbers, each one identical to the first comprising just the values one to nine.

1 2 3 4 5 6 7 8 9

This was intriguing. Nine digits running through everything. My moment of idleness vanished and I became excited, asking the question: how had the number 9 jumped out at me so easily, from out of these figures? The answers seemed to lie not in the numbers themselves but under the surface – where undercurrents left a different trace, revealing hidden patterns to a secret eye.

So with the number nine as inspiration, believing in its royal and creative values, I decided to use this shorthand to get me into the heart of the riddle: What were the fixed points in the wind? And what did number nine have to do with this; was nine the end of the line or something else, at the fixed centre of things?

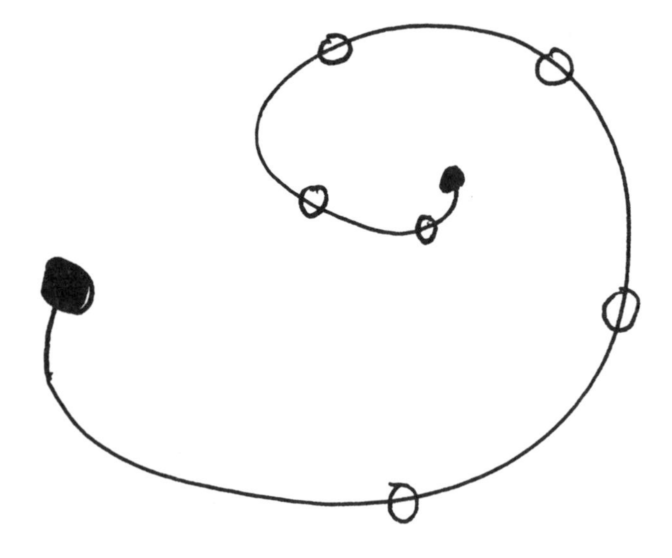

The Sigma Code

Each and every number has a secret number. Buried beneath the surface, hidden within the construction is another mark, a secret code that defines the original number. Like the elements in chemistry, like the alphabet in language, it is a primary classification from which all else flows. It is an imprint.

It is a digital code with only nine values.

1 2 3 4 5 6 7 8 9

Each number reduces to just one such value when its integers are summed up.

32	*has the hidden number 5*	*for*	$3+2 = 5$
81	*has the hidden number 9*	*for*	$8+1 = 9$
100	*has the hidden number 1*	*for*	$1+0+0 = 1$

Because secret numbers are derived by an additive process I call them sigma numbers, using the Greek letter Σ (sigma) which in time-honoured tradition in mathematics stands for adding things up.

$\Sigma 32 = 5$
$\Sigma 81 = 9$
$\Sigma 100 = 1 \quad (1+0+0=1)$
$\Sigma 148 = 4 \quad (1+4+8=13;\ 1+3=4)$

What happens with a big number?

If the digits of 1035 721 are added up
we get $\Sigma 1035\,721 = 1$

$$
\begin{aligned}
& 1+0+3+5+7+2+1 = 19 \\
\textit{and} \quad & 1+9 = 10 \\
\textit{and} \quad & 1+0 = 1
\end{aligned}
$$

The sigma value of a number
is the ultimate essence of a number.
It is the hidden mark which lurks
within the greater construction;
in this sense it is a primary code,
a blueprint.

Mental additions can be done rapidly due to a strange property of the sigma code. If there is a 9 in the number it can be left out. Moreover, as one adds up the digits, every time the sum comes to nine or a multiple of nine these accumulations or partial sums can be left out and the addition continued.

Examples:

$\Sigma 392 = 5$

I get the same answer if I add three and two, dropping the nine, as when I add up all the numbers three, nine and two, and get fourteen, for 1 + 4 = 5.

Suppose I want to add the digits of the number:

4 9 5 3 4 2 6 1 9

If I start from the left, I leave out the first 9 and add; 4 + 5. But this sums to 9 so I leave it and carry on; 3 + 4 + 2 which also sums to 9, so I drop this as well and carry on; 6 + 1 = 7; and I leave out the last 9. The sequence of short cut additions give a final answer of 7.

$\Sigma\ 4\,9\,5\,3\,4\,2\,6\,1\,9 = 7$

When adding up the digits, one can start anywhere. Seize on the pattern that gives the simplest, quickest addition, looking out for the dropping of nines. Very quickly the reader will become adept at really fast addition, anticipating which numbers add to nine and can be left out without affecting the final answer.

Number Trace

If we want to know what the sigma value of 16 x 253 is we only need to multiply their individual sigma code values to get the number we want.

$\Sigma 16 = 7 \qquad \Sigma 253 = \Sigma 10 = 1$

The answer is one times seven which is seven and this checks out with the longer calculation.

$\Sigma(16 \times 253) = \Sigma 4048 = \Sigma 16 = 7$

You see how powerful the method is to check calculations. If there are any errors in the calculations the code will reveal it, because numbers cannot shake off their blueprint.

The sigma code leaves a permanent trace.

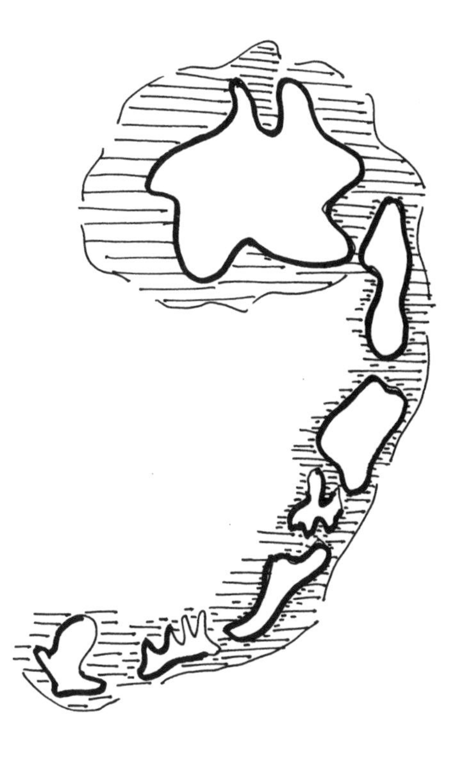

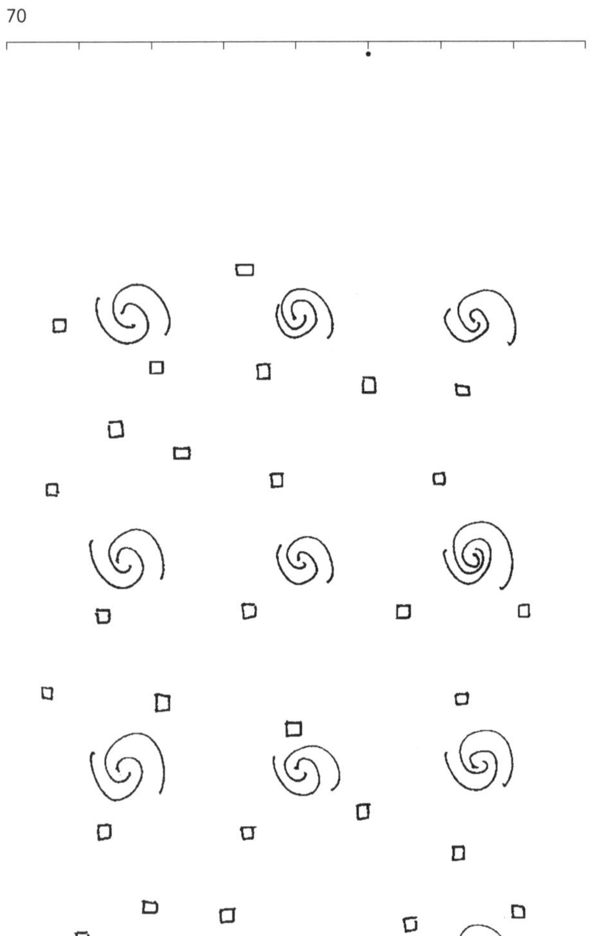

Four Precious Mirrors

Addition: Clue 5

Subtraction: Clue 6

Multiplication: Clue 7

Division: Clue 8

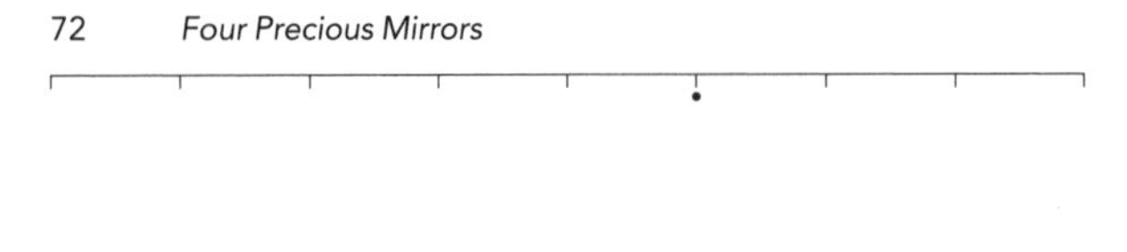

Four Precious Mirrors

Armed with the code we go on to look into the precious mirrors of arithmetic, the four infinite planes of adding, subtracting, dividing and multiplying, that manipulate all numbers.

The sigma code lays down a trace of how numbers work in secret, we can peer into their basic patterns. But it is only when these patterns are seen as a whole that the beauty of the code is revealed and the astonishing truth of number nine at the heart of our number system.

(And here we must thank zero for even allowing the sigma code to exist. For without zero the process of reduction would not work. Adding digits to pare down to a single digit is dependent on our unique placement value system of units, tens, hundreds, etc. Zero allows the code to be bound within the range 1–9).

Clue 5: Addition

When 9 or a multiple of 9 is added to a number, the answer has the same sigma value as the original number.

$33 + 9 = 42$

Written down more formally according to the sigma code:

$\Sigma 33 = \Sigma 42$

$(3 + 3) = (4 + 2)$

The addition of 9 leaves no trace! But try this with another number and it does not work. Add, say, seven to thirty-three:

$33 + 7 = 40$

As before the digits 33 add up to six, (3 + 3); but the corresponding answer of forty adds up to (4 + 0) giving four. And six does not equal four.

Other examples:

$132 + 9 = 141$
$44 + 18 = 62$
$11 + 36 = 47$

In each of the above examples the number on the left of the addition has the same sum of digits as the corresponding answer on the right hand side. Nine seems to change nothing under the sigma code.

That is:

$\Sigma 132 = 6 = \Sigma 141$
$\Sigma 44 = 8 = \Sigma 62$
$\Sigma 11 = 2 = \Sigma 47 \qquad (4 + 7 = 11; \quad 1 + 1 = 2)$

This indeed is a strange situation. In the world of arithmetic, addition makes things grow and the outcome is expected to be bigger than any one of the starting numbers. But when the hidden trace of numbers is investigated something else happens. There is no change at all when nine or one of its multiples is added to a number. In secret, as it were, the original number is allowed to pass unchanged. If any number other than nine is used in the addition the hidden trace is altered.

Consider one more example, of say six being added to forty-four. The answer, fifty, has digits that do not sum up to the digits of the number forty four.

For example:

$$\begin{array}{ccc} 44 & +6 = & 50 \\ 4+4 & \neq & 5+0 \end{array}$$

But if 9 had been added then

$$\begin{array}{ccc} 44 & +9 = & 53 \\ 4+4 & = & 5+3 \end{array}$$

Again number nine leaves no trace. In the first precious mirror of addition, 9 has no reflection; it is a ghost!

What happens when nine is added to itself?

$$\begin{array}{ll} 9+9 & = 18 \\ 9+9+9 & = 27 \\ 9+9+9+9 & = 36 \\ 9+9+9+9+9 & = 45 \end{array}$$

etc.

This, of course, is also the multiplication table for nine.

In each case the digits add up to 9. However many 9s are added, the outcome is the same; the answer always sums up to nine. Nothing changes. Nothing alters. Only the fixed mark of 9 remains.

$9 + (12 \times 9) = 117 \quad and \quad (1 + 1 + 7 = 9)$

This is quite amazing. It is as if nine is stationary and does not move at all!

The multiplication table of number nine also gives curious reflections.

$9 \times 1 =$ **09**	**90** $= 10 \times 9$
$9 \times 2 =$ **18**	**81** $= 9 \times 9$
$9 \times 3 =$ **27**	**72** $= 8 \times 9$
$9 \times 4 =$ **36**	**63** $= 7 \times 9$
$9 \times 5 =$ **45**	**54** $= 6 \times 9$

The first five products are reflected in the next five products as if a mirror runs between them.

Clue 6: Subtracting Numbers

Subtracting 9 or a multiple of 9 from a number does not alter the sigma value of that number – its secret value does not change.

$$44 - 18 = 26$$
$$132 - 9 = 123$$
$$1089 - 27 = 1062$$

The reader may verify that the sum of the digits of the original number on the left hand side prior to subtraction is the same as the sum of the digits in the answer. The reduction and shrinking has not been affected by nine.

But if numbers whose digits do not add up to nine are subtracted from these numbers, say by even the difference of one in the value of the number being subtracted, then the secret value of the answer is quite different to that of the original number.

$$44 - 17 = 27 \qquad \Sigma 44 \neq \Sigma 27$$
$$132 - 10 = 122 \qquad \Sigma 132 \neq \Sigma 122$$
$$1089 - 26 = 1063 \qquad \Sigma 1089 \neq \Sigma 1063$$

Summary:

Upon addition and subtraction number 9 leaves no trace. On the surface of appearances in arithmetic, everything changes, every addition or subtraction has a unique value. But under the surface, when tracked by the sigma code, it is as if arithmetic freezes in time when the character of nine enters the scene. Nothing is changed by plus or minus. In the mirrors of addition and subtraction nine is an unseen ray.

How different then the behaviour of nine on multiplication. From being an unseen ghost it turns now into an attacking force, eating up the characters of other numbers, pushing itself into a position of permanent domination. Upon multiplication, as revealed by the sigma code, number 9 is like a virus, an invincible marauder.

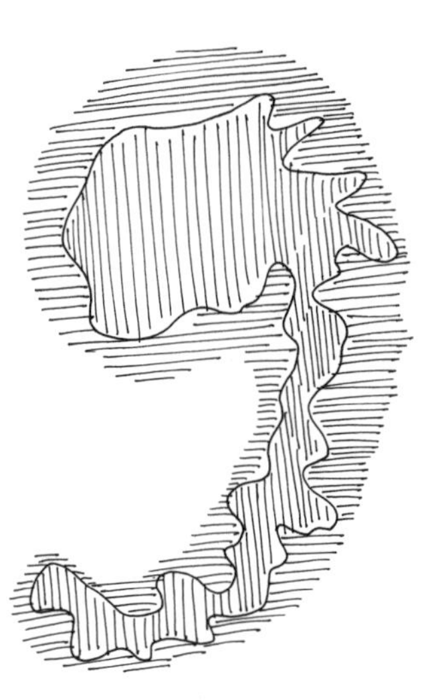

Clue 7: Multiplication

When a number is multiplied by 9 or a multiple of 9 then the digits of the product always add up to 9.

$$7 \times \mathbf{9} = 63 \qquad \Sigma 63 = 9$$
$$4 \times \mathbf{45} = 180 \qquad \Sigma 180 = 9$$
$$11 \times \mathbf{72} = 792 \qquad \Sigma 792 = 18\ (1 + 8 = 9)$$

Once a multiplier has the secret value 9, it seems to stamp its authority onto any multiplication and takes over the identity of the numbers it multiplies.

Again, as with addition and subtraction, it is nine that has the unique character, not any other number. With other multipliers, for example, the product of multiplication does not have the same sigma value as the multiplier.

$$7 \times \mathbf{8} = 56 \qquad \Sigma 56 \neq 8$$
$$4 \times \mathbf{43} = 172 \qquad \Sigma 172 \neq \Sigma 43\ (= 7)$$
$$11 \times \mathbf{74} = 814 \qquad \Sigma 814 \neq \Sigma 74\ (= 2)$$

If a number is multiplied by a series of numbers and 9 is one of them, then the sum of the final product again is 9.

$7 \times (2 \times 3 \times 4 \times 9 \times 6) = 9072$

Σ 9072 adds up to 18, and 1 + 8 = 9.

On multiplication, nine emerges as a mighty force – quite different to its unseen behaviour in addition and subtraction. There is no change to the sigma code value of a number when 9 is added or subtracted but upon multiplication number nine gains control. Why?

To summarise then what arithmetic has shown us so far, if all numbers are reduced to a secret value held by the sigma code:

$$\Sigma\, \textit{number} + \Sigma\, 9 = \Sigma\, \textit{number}$$

$$\Sigma\, \textit{number} - \Sigma\, 9 = \Sigma\, \textit{number}$$

$$\Sigma\, \textit{number} \times \Sigma\, 9 = \Sigma\, 9$$

Nine behaves identically in two precious mirrors – for it gives no sign of its existence in either addition or subtraction. In the third mirror, multiplication, it thrusts itself forward in a dominant mode. One would expect then in the reflection of multiplication, viz division, that in this fourth mirror, number nine would feature strongly. And it does.

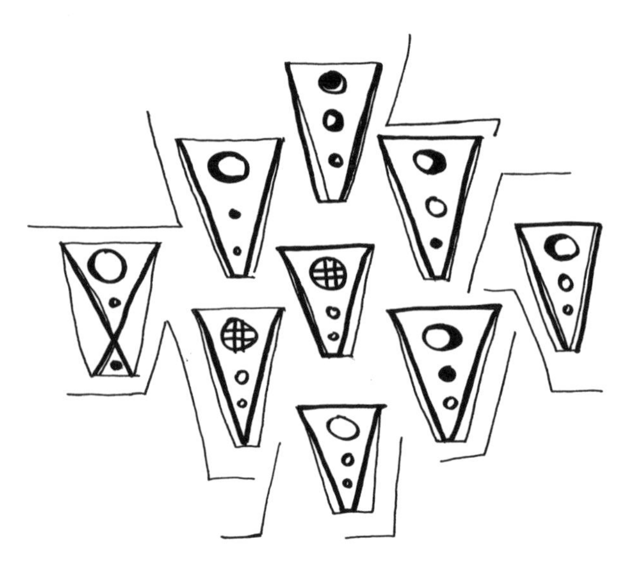

Clue 8: Division – the last of four mirrors

If a number is divided by 9, the remainder has the same value as the sum of the digits of the original number.

$62 \div 9$

When 62 is divided by 9 the remainder is 8, identical to the sum of the digits of the original number 6 + 2 = 8

Example:
When 439 is divided by 9 the remainder is 7

$\Sigma 439 = 4 + 3 + 9 = 16; \qquad 1 + 6 = 7$

If a number is exactly divisible by 9 giving a remainder of zero, then the sum of its digits is 9 or a multiple of 9.

Example:
432 sums up as 4 + 3 + 2 = 9; it is therefore exactly divisible by nine.

$432 = 9 \times 48$

Similarly 288, 126, 8 919, 112 923 are numbers exactly divisible by 9. In each case the sigma value of the number is nine. Zero seems to be related in some way to Σ9.

Question?

Is nine something static, another number in line after seven and eight, or do we think by now that it is something more wilful and dynamic, vanishing and being seen as it pleases?

If we had followed additions and multiplications up on the surface of arithmetic we would have noticed nothing, just numbers. Patterns would have been hard to find. But when marked by the special code, numbers do not escape their blueprint. They leave a trail and it is here, in the hidden world of numbers, where the full richness of the patterns of number nine are revealed.

Nine and Zero

Under the sigma code, in addition and subtraction nine has no effect, just as if zero is added or subtracted to a number. In multiplication nine takes over the identity of a number, just as zero does. And in normal arithmetic, exact division by a number gives the remainder zero – according to the sigma code if a number can be divided by 9 exactly its digits add up to nine.

There seems to be a strange affinity between zero and $\Sigma 9$.

Clue 9: Multiplication Tables

Most of the clues are now with us. But something is still missing, one more clue is needed to unravel the mystery. For the ninth clue it is best to take a closer look at multiplication, where nine revealed itself in the most direct and forced manner so far. A search into these products should be fertile ground for more revelation.

But can we find something more interesting in these products than just their jumping and ever increasing size?

Would it not be nice to find a hidden pattern at work that would make learning the times table more fun, and not the labour so many have to learn the tables by rote, in a dull and uncomprehending manner?

The 2× table

Let's keep it simple and start with the easiest multiplication, that of the 2× table which gives jumps in twos:

2 – 4 – 6 – 8 – 10 – 12 – 14 – 16 – 18 – 20 – 22 – 24 – 26 ...

By adding up the digits this series of numbers can be transformed into the sigma code as follows:

2 4 6 8 1 3 5 7 9 2 4 6 8...

Here is a surprise! Instead of something that grows forever in jumps of two the sigma code shows a repeating pattern of nine values:

2 4 6 8 1 3 5 7 9

However far the multiplication goes the hidden pattern repeats itself. Written below, as a further example, are the products from 266 (2 × 133) to 300 (2 × 150). The corresponding code values are in brackets.

(Remember 266 has the sigma value 5; for 2 + 6 + 6 = 14; and 1 + 4 = 5).

266	*(5)*	
268	*(7)*	
270	*(9)*	
272	***(2)***	
274	***(4)***	
276	***(6)***	
278	***(8)***	
280	***(1)***	The repeating sequence
282	***(3)***	
284	***(5)***	
286	***(7)***	
288	***(9)***	
290	*(2)*	
292	*(4)*	
294	*(6)*	
296	*(8)*	
298	*(1)*	
300	*(3)*	

Like some kind of inner mantra the pattern is repeated over and over again; a self stabiliser in the jumping growths of multiplications. The 2× table has only one composition when seen through the x-ray of the code. Its blueprint is:

2 4 6 8 1 3 5 7 9

Multiplication by 3× table

The products of the first nine values are:

3	*6*	*9*	*12*	*15*	*18*	*21*	*24*	*27*

The corresponding secret values or sigma code values are:

3	*6*	*9*	*3*	*6*	*9*	*3*	*6*	*9*

However far the multiplications go the hidden pattern is indelible, marking the products in steps of ③–⑥–⑨, the stepping stones of nine.

From these results it looks like only the first nine values of a multiplication table need to be studied to find a hidden pattern. (The character of nine dominates again).

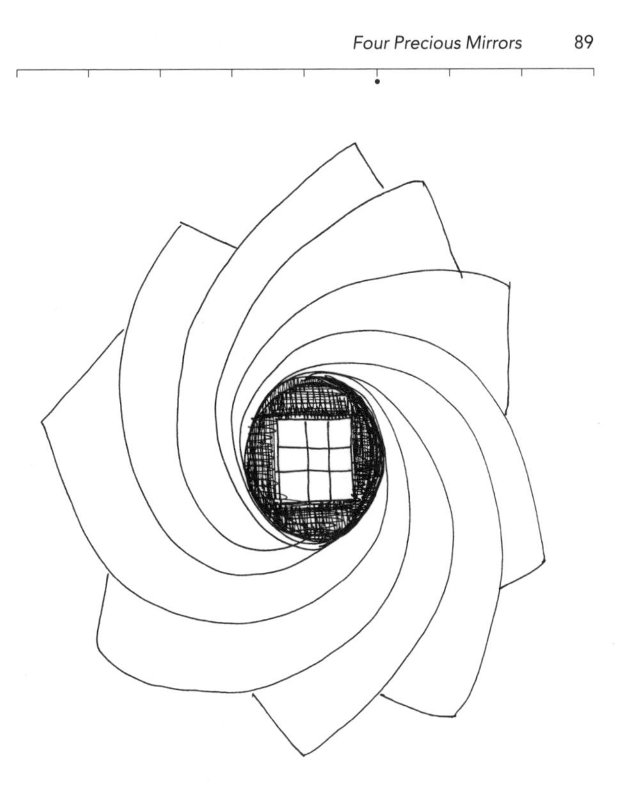

It is time now to look at the entire table of multiplications along with the corresponding code values of their products to see what else we can find. It is here that we first come across the astonishing secret of the numbers 1 – 9 in the sigma code. They are twinned!

Figure 1:
Multiplication Table

	×1	×2	×3	×4	×5	×6	×7	×8	×9
1	1	2	3	4	5	6	7	8	9
2	2	4	6	8	10	12	14	16	18
3	3	6	9	12	15	18	21	24	27
4	4	8	12	16	20	24	28	32	36
5	5	10	15	20	25	30	35	40	45
6	6	12	18	24	30	36	42	48	54
7	7	14	21	28	35	42	49	56	63
8	8	16	24	32	40	48	56	64	72
9	9	18	27	36	45	54	63	72	81

Figure 2:
Sigma code values of the multiplication table

	×1	×2	×3	×4	×5	×6	×7	×8	×9
1	1	2	3	4	5	6	7	8	9
2	2	4	6	8	1	3	5	7	9
3	3	6	9	3	6	9	3	6	9
4	4	8	3	7	2	6	1	5	9
5	5	1	6	2	7	3	8	4	9
6	6	3	9	6	3	9	6	3	9
7	7	5	3	1	8	6	4	2	9
8	8	7	6	5	4	3	2	1	9
9	9	9	9	9	9	9	9	9	9

Figure 1 contains the multiplication tables up to ×9 for the numbers 1–9. Figure 2 tables the corresponding sigma values of these products. We need not go beyond 9 × 9, for ×10 or ×14 say, are the same as multiplying respectively by 1 ($\Sigma\,10 = 1 + 0 = 1$) or 5 ($\Sigma\,14 = 1 + 4 = 5$).

Unlike the multiplication tables which go on and on, *all multiplications are condensed in the sigma code sequences shown in figure 2.* Within that table is the blueprint for all doubling, tripling, quadrupling, etc.

Every possible multiplication is reduced to the matrix of nine by nine, just eighty one values. And yes, '81' as 8 + 1 adds back to 9! Nine permeates everything.

But the startling discovery is that there are interrelationships between the values of the sigma code itself. The code is not a set of independent values as we might have first thought. I had been interested in tracking just the number nine, but now a more intricate and surprising world of interconnection is revealed.

The blueprint for each 'times' table has a matching reverse! Multiplication under the sigma code shows that except for 9, the basic numbers 1–8 are twinned with each other. Each sequence is the opposite of the other, except for 9, which seemingly acts as an end stop.

Nine refuses to act as the other numbers do, having no opposite. Instead the number stands aloof and acts as a border to each array, stopping any flowing outwards and inwards of a sequence. Number nine is the end of the line watchdog – keeping a strict limit to multiplications under a secret code. Here are the 'twins' from the table in figure 2:

4×	4	8	3	7	2	6	1	5	9
5×	5	1	6	2	7	3	8	4	9

3×	3	6	9	3	6	9	3	6	9
6×	6	3	9	6	3	9	6	3	9

2×	2	4	6	8	1	3	5	7	9
7×	7	5	3	1	8	6	4	2	9

1×	1	2	3	4	5	6	7	8	9
8×	8	7	6	5	4	3	2	1	9

If the numbers vertically above each other are added, they sum up to 9!

Number 9 'locks' in the sequences. Nine refuses to act as the other numbers do, having no opposite.

It is interesting to map the values of 9 in the sigma multiplication table.

• • • • • • • • 9

• • • • • • • • 9

• • 9 • • 9 • • 9

• • • • • • • • 9

• • • • • • • • 9

• • 9 • • 9 • • 9

• • • • • • • • 9

• • • • • • • • 9

9 9 9 9 9 9 9 9 9

Number 9 features as a limiting value, as a border; but there is also the symmetry of 9, marking out the corners of a square in the middle of the table.

Is number nine the end of the line, a border, or a centre? Here it shows a dual personality, both as centre and border.

What pattern is buried within the code?

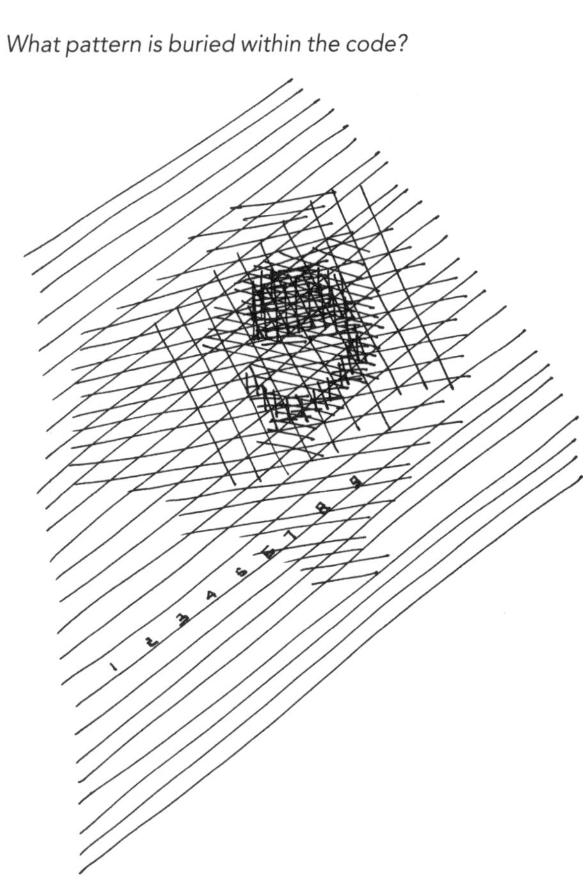

Summary of the nine clues:

Clue 1:
The ninth point in the directions of the wind is at the centre. It is the point from which all else flows.

Clue 2:
In myths and legends nine has the quality of initiation and beginning. So nine is not the end of the line, but more a jumping-off point, a beginning, a point of origin.

Clue 3:
Nine is closely allied to reversals as in the trick with three figure numbers. Nine is some kind of mirror.

Clue 4:
The sigma code, the hidden blueprint of all numbers, has nine values. The nature of nine is the code itself. (It exists, in other words, to serve nine.) Like a selfish gene in the chain the character of nine dominates the basic digital code of all numbers.

Clue 5:
In addition number nine leaves no trace.

Clue 6:
In subtraction number nine leaves no trace.

Clue 7:
Multiplication by nine always ends in nine. In this case imagine a causeway of stones; if we skip in twos or threes or fours or whatever, but repeating it nine times (the equivalent of multiplying by nine), then we always land on a stone marked by nine.

Clue 8:
A number upon division by nine leaves a remainder equal to the sum of its digits. If the digits of the original number add up to nine, then the number is exactly divisible by nine. $\Sigma 9$ has an echo of zero to it.

Clue 9:
The multiplication tables, as revealed by the sigma code, have in-built symmetries between numbers 1/8, 2/7, 3/6, 4/5. Number nine acts as both border and centre in the secret map we make of all our multiplications.

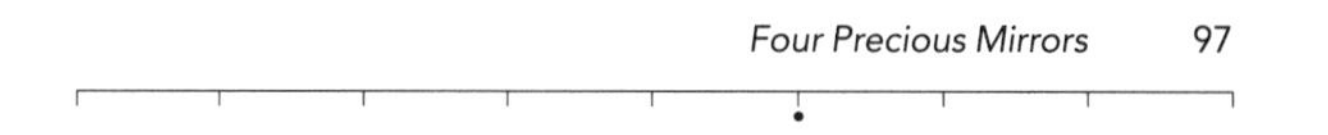

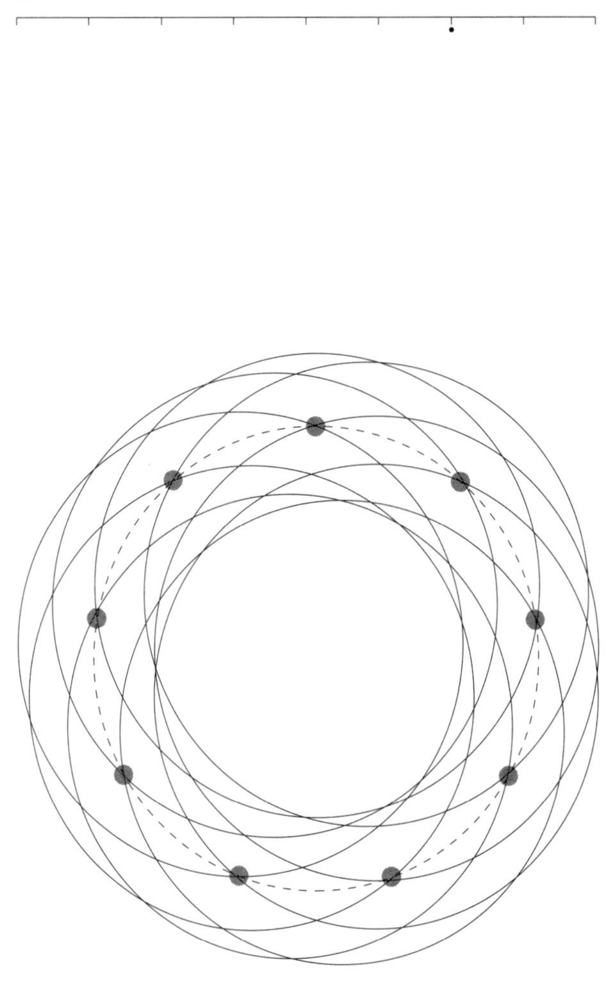

Mandala

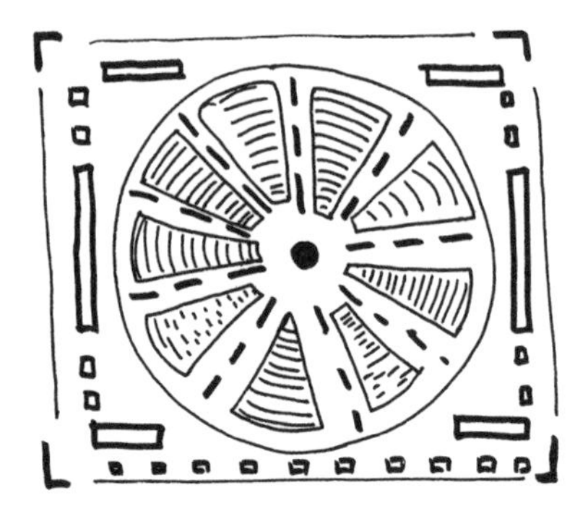

Enjil's strategy and first solution

All these facts and questions passed through the young master's mind. He had done the work and contemplated all the clues. The search had been exciting, it revealed a world of hidden negotiations that he had not dreamt of. The strategy of using secret numbers was powerful – it gave ready access to larger calculations and the noting of hidden rhythms.

But how nine appeared or disappeared in arithmetic when tracked by the sigma code was a puzzle. Something very special was going on and Enjil knew it needed a unique solution, something that would have great beauty but also something the children of the village could play with. Why not? Why should the solutions be open to only the few, the so-called experts?

He climbed up the mountain behind the Academy and bathed in cold water to refresh his mind. He put on clean clothes next to his skin.

He meditated hard upon the clues, all nine of them and thought afresh again. But there was still no answer. There were intuitions; but the facts kept confusing him, they pulled in different directions. The pattern he sought was hard to catch. It seemed to move and reflect and have motion, and yet it had the habit of not moving, being fixed like a totem pole in the centre of a village.

Enjil became more agitated, afraid he had abandoned more high-minded algebra in favour of child-like curiosities, a wilfulness that would now cost him dearly and leave him empty-handed before the critical and cutting eye of the Elders. And the Examination would not wait; there were only two days left. He knew he had it in him to solve the riddle but if only he could see the direct path to it.

What or where was this fixed point in the wind?

When his mind became tangled into a knot as it was now he had learnt to cut out the reasoning parts and let emptiness flow in; and as the sun went down Enjil entered this phase of being an empty vessel. Night fell.

The moon came up and somehow the surrounding blackness helped his mind grow vacant of all facts and numbers, and he listened to the rustling leaves as the moving air silently put the first patterns into his head. He saw numbers coming from several parts and going to many parts, just like the wind.

Any number, like seventy-two say, stood for an infinity of things, from ants to seeds, to cows, plants and constellations. But number seventy-two itself could be made up in many ways by other numbers. Even unity, the number one, could be composed a million ways by the combination of different fractions; so a number had motion passing through it, like a river flowing. Each number was part of other numbers – for one instant we catch the essence as the pool eddies and forms a stillness, a stationary point, before dissolving again. Perhaps numbers are poles of frozen abstraction, quiet points in an otherwise constant movement? But these thoughts took him into the philosophy of numbers and that was something else to debate with the Elders, on another day. Now he needed to open his mind to find the pattern he sought, for like with most things, he knew it was Pattern and not Number that would be the answer to his quest.

The moon came out in all its brightness. It was full. He gazed at the disc wondering if there really was a woman spirit who lived in the illuminating light, cold and white when she was inspired, and warm and yellow when her spirit moved tender hearts to open like giant, magic mushrooms?

Enjil closed his eyes around the moon.

A shape came into them. He tested it against the clues. The numbers rushed around and fitted. It worked! As he extended the mental calculations the shape grew and made an even better fit. Each clue was tested. Several patterns overlaid his eyes in a beautiful symmetry. Oh the joy of discovery! Then there was more; in a sudden twist the pattern became chaotic and surprisingly even more beautiful to his eye. He knew he had it! It was what he had looked for all along. The search was over.

And Enjil took the bright visions he saw into the drawings he prepared for his thesis. Whether the Elders passed or failed him he did not care; what mattered was the joy he found at unlocking the secret world of numbers, hidden way below the surface of what was arithmetic. Moreover, he had also solved the riddle the woman had given him. He had found the fixed points in the wind, not just one as he had thought, but all nine of them!

The Pattern

Instead of imagining the sigma code as a straight line of values, everything falls into place if the code is arranged in a magic circle.

Number nine, the royal nine, takes pride of place at the top; everything flows from there, clockwise, as the circle divides into nine sections.

Each section of the circle subtends forty degrees at the centre.

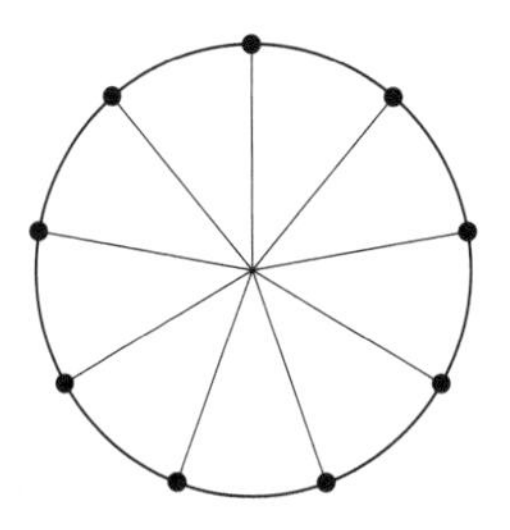

The numbers across the circle sum up to nine:

1 + 8, 2 + 7, 3 + 6, 4 + 5

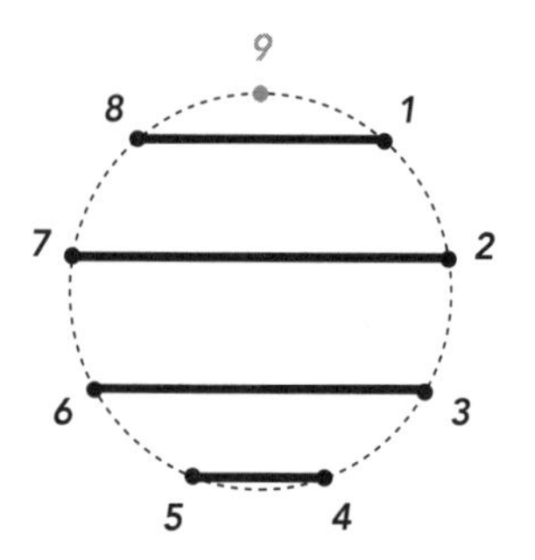

Reading across the circle and pairing numbers first right-left and then left-right we find the nine times table:

18 27 36 45 54 63 72 81

This is like imagining a vertical line going through nine and acting as a mirror

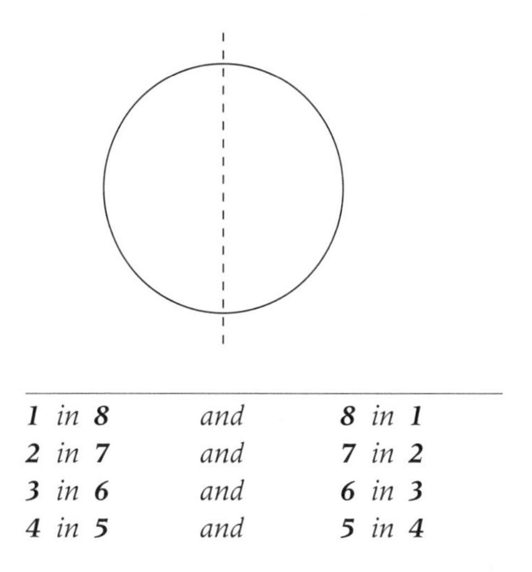

1 *in* ***8***	*and*	***8*** *in* ***1***
2 *in* ***7***	*and*	***7*** *in* ***2***
3 *in* ***6***	*and*	***6*** *in* ***3***
4 *in* ***5***	*and*	***5*** *in* ***4***

exactly reflecting the pairings revealed by the sigma code in multiplication, where one sequence is the exact reverse of the other except for 9. That is because 9 *is the mirror plane; it has no reflection*. It is quite beautiful that the strange patterns of the multiplication table in code are so easily explained as a natural consequence of the number 9 arranging its component parts around a circle, making the twin pairings inevitable.

In a vertical direction, the numbers one above the other on the left add up to 13 and on the right to 5. But 13 reduces to 1 + 3 = 4, so we have in the circle:

left-handedness as 4
right-handedness as 5; *and 4 + 5 = 9*

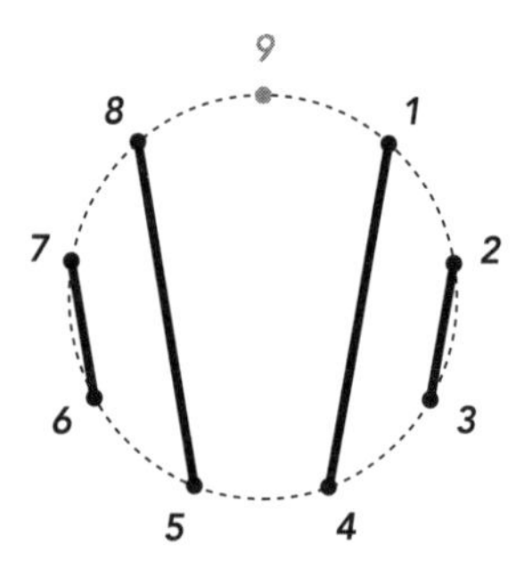

Remember that each number represents a code value, a summation of the digits of numbers on the surface of arithmetic. Here in the circle of nine, we swim through the undercurrents of numbers. Calculations run through and around the circle.

The sigma circle also explains the puzzles we encountered in the four mirrors of arithmetic. Without recourse to the mathematics of algebraic proofs, the circle simply draws a picture around each clue.

We move around the circle to find the answers. And like a good detective story, after the puzzle is set we must have an explanation. So let's begin with number tricks and reversals.

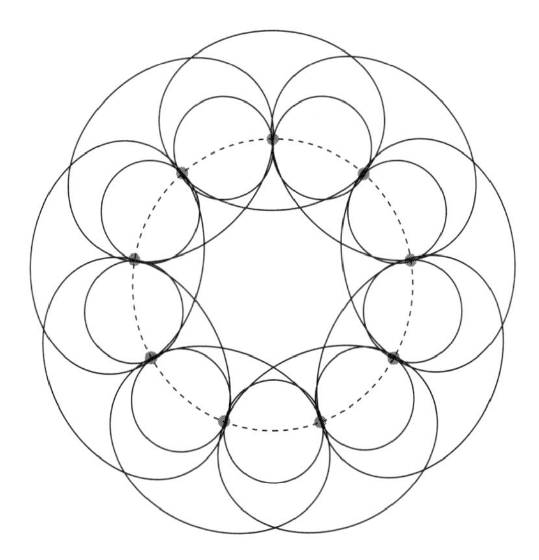

Reversals and Party Tricks: Clue 3

Take a two digit number, say number 31. Then after reversal and subtraction we get the number 18.

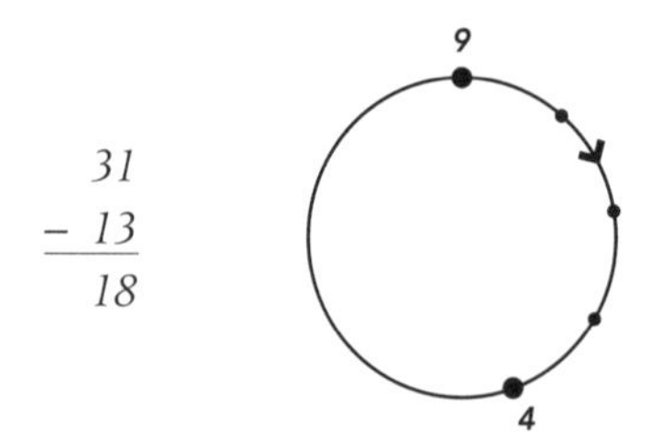

Now 31 has the sigma value 4 (3 + 1 = 4). So I travel along the circle starting from 9, four stops, to arrive at 4.

When 31 is reversed to the number 13, the new number has the same sigma value, for 1 + 3 still equals 4. Subtraction means going into reverse on the sigma circle, so I go back four stops. I arrive back at 9. This means that any number, once reversed and subtracted, will always give an answer whose digits will add up to 9.

Now, after reversal, the answer 18 has the sigma value 9. If the answer is reversed yet again, say to 81, it still has the sigma value 9, for 8 + 1 = 9. Nothing changes.

We could go on reversing and adding, or reversing and subtracting, it does not matter. We travel a full 9 stops around the circle, backwards or forwards, always arriving back at 9. What matters is that first reversal which brings us back to point 9 on the sigma circle. Thereafter, subsequent reversals keep us at the same point.

Going on to three digit numbers,

Number	*781*	$\Sigma\, 781 = 7$
Reverse	*– 187*	$\Sigma\, 187 = 7$
	594	$\Sigma\, 594 = 9$
Reverse	*495*	$\Sigma\, 495 = 9$
Add	*1089*	$\Sigma\, 1089 = 9$

The original number of 781 adds up to 7. Go to point 7 on the circle. Reversing and subtracting moves us back seven places (reversing does not change the sum of digits) and we reach 9. This point corresponds in the calculation to the number 594 – whose sigma value must also be 9. And this indeed is the case.

Now if 594 is reversed to 495 the sigma value is unchanged, as $\Sigma\, 495$ is the same as $\Sigma\, 594 = \Sigma\, 9$.

We are back at 9, the point we had reached in the first place. And so on . . . . We can extend this game beyond anything the old number tricksters imagined. We can use any numbers, small or big, (even negatives or decimals if we are careful to follow the correct rules). We can reverse and re-reverse any amount of times; the final answer will always sum to nine.

The explanation of Clues 5 and 6 now become obvious.

Clues 5 and 6

No traces in the mirrors of addition and subtraction

Adding nine to a number leaves no trace, for adding 9 is to travel a full nine stops around the circle to come back to where we were in the first place.

$$\begin{array}{r} 132 \\ +\quad 9 \\ \hline 141 \end{array} \qquad \begin{array}{l} \Sigma\,132 = 6 \\ \\ \Sigma\,141 = 6 \end{array}$$

The number we add need not be 9. The digits of the number need only add up to 9, like say 171. Then the same result happens.

$$\begin{array}{r} 132 \\ +\ 171 \\ \hline 303 \end{array} \qquad \begin{array}{l} \Sigma\,132 = 6 \\ \\ \Sigma\,303 = 6 \end{array}$$

Adding Σ 9 is to travel around the circle a full nine stops.

Similarly with subtraction:

132	Σ 132 = 6
− 9	
123	Σ 123 = 6

Subtraction acts just like addition. Taking away 9 from a number is to go back round the circle, a full nine stops, arriving at the same point. I have used small numbers to keep the sums simple but the rule holds good for any number, even large numbers.

83426813	Σ (8)
− 31253616	Σ (9)
52173197	Σ (8)

The first large number has digits that add up to eight; its sigma value is 8. The number to be subtracted has the sigma value 9 and thus will not affect the outcome, for we travel round the circle coming back to 8. The answer therefore should have the sigma value 8, which is the case.

Multiplication by Nine: Clue 7

$2 \times 9 = 18 \qquad \Sigma 18 = 9$

Remember the numbers move from the fixed point of 9. To multiply nine times we must take 9 steps of 2. The first step is obtained by moving 2 along from 9; and we arrive at 2. Then eight other similar moves in twos will finally reach 9 again. It is easy to see that with any number this will happen, because there are nine divisions to the circle. Try it out.

Example:

$3 \times 9 = 27 \qquad \Sigma 27 = 9$

Three times one takes us to point 3 on the sigma circle because 3 moves out of 9 to take its place at position 3 on the sigma circle. Then eight other jumps of 3, means we have taken $3 \times (1 + 8)$ jumps or 3×9 jumps; and we find ourselves at the top of the circle, at the pole position again.

Σ 9 is the fixed point for all growth in nines.

Division: Clue 8

Number 481 resides at point 4 on the sigma circle, and when divided by 9 its remainder is, surprisingly, also four. Similarly, when say 53 is divided by nine, the remainder is eight, and number 53 resides at point 8 on the sigma circle ($\Sigma\, 53 = 5 + 3 = 8$).

The sigma value of a number is its remainder when divided by nine.

But if the sigma value of a number is 9, like say 261, then under division by 9 there will be no remainder.

Σ 9 corresponds to remainder zero.

Numbers such as 621, 432, 411 111, each add up to nine as well. Such numbers whose sigma values are 9, are privileged, they reside at the top of the circle, at the pole position. They are on a mirror axis and do not move, as if anchored to a zero point.

In the dynamics of the sigma circle Σ 9 must be seen as fixed. The other values move out from Σ 9 to take their place in the circle. We travel to the other numbers but we do not move when we are at the position of the number nine. It is as if the other numbers turn and spin away or towards nine; but at the top, at the pole, the ninth spot does not move.

Number Shapes from the Sigma Circle

But what are the shapes of the numbers themselves as seen by the secret code in multiplication?

The products of multiplication according to the sigma code are:

×1	1	2	3	4	5	6	7	8	9
×2	2	4	6	8	1	3	5	7	9
×3	3	6	9	3	6	9	3	6	9
×4	4	8	3	7	2	6	1	5	9
×5	5	1	6	2	7	3	8	4	9
×6	6	3	9	6	3	9	6	3	9
×7	7	5	3	1	8	6	4	2	9
×8	8	7	6	5	4	3	2	1	9
×9	9	9	9	9	9	9	9	9	9

If each horizontal row is now plotted on the sigma circle and the points joined together, we get the shapes of the numbers 1–9 in multiplication as seen on the diagram on the right. This diagram is repeated on the following page to give clarity to the comments that follow.

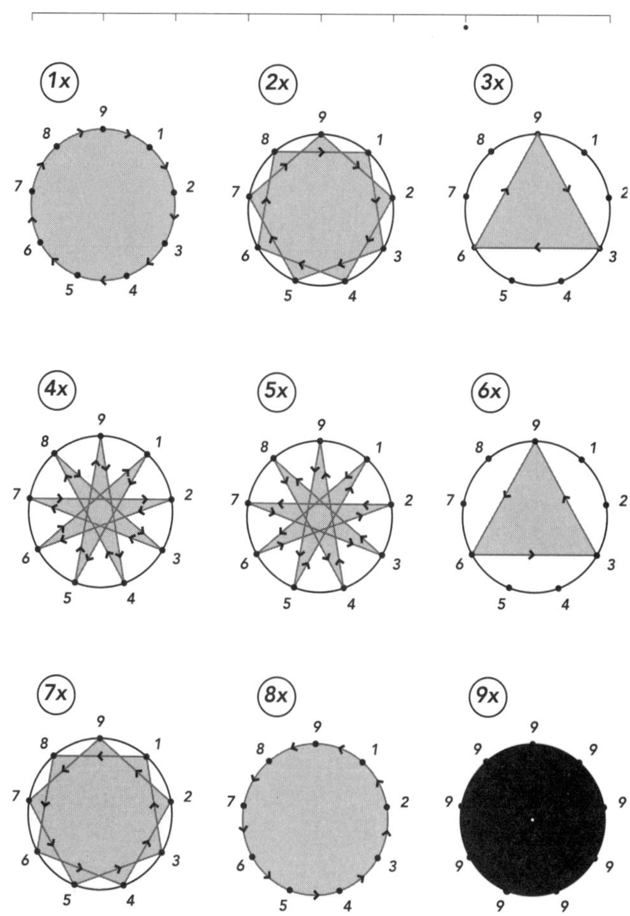
1x
9 1 2 3 4 5 6 7 8
2x
9 1 2 3 4 5 6 7 8
3x
9 1 2 3 4 5 6 7 8
4x
9 1 2 3 4 5 6 7 8
5x
9 1 2 3 4 5 6 7 8
6x
9 1 2 3 4 5 6 7 8
7x
9 1 2 3 4 5 6 7 8
8x
9 1 2 3 4 5 6 7 8
9x
9 9 9 9 9 9 9 9 9

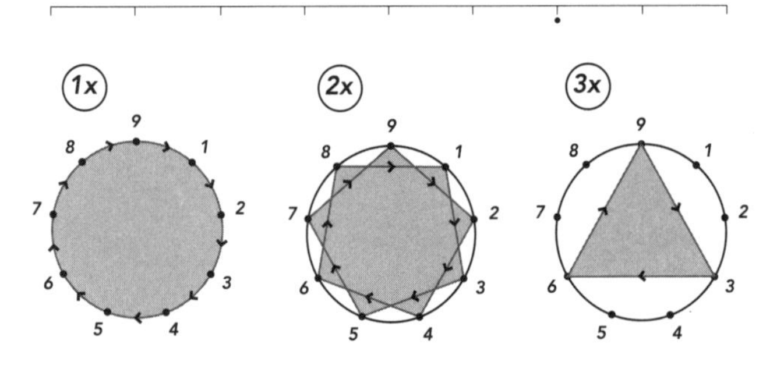
1x
9
8
1
7
2
6
3
5
4
2x
9
8
1
7
2
6
3
5
4
3x
9
8
1
7
2
6
3
5
4

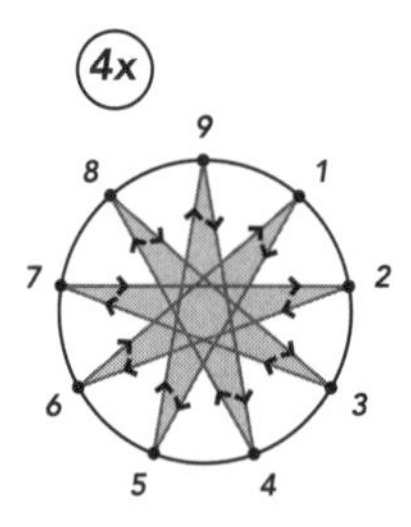
4x
9
8
1
7
2
6
3
5
4

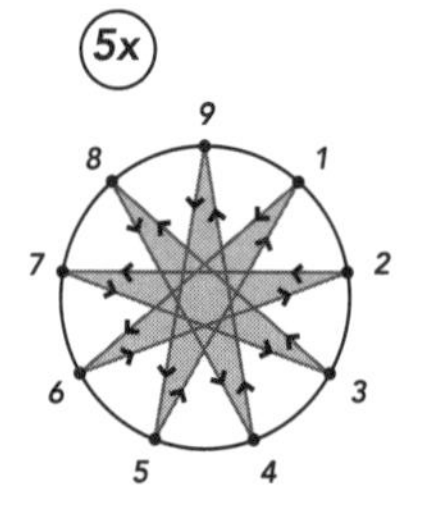
5x
9
8
1
7
2
6
3
5
4

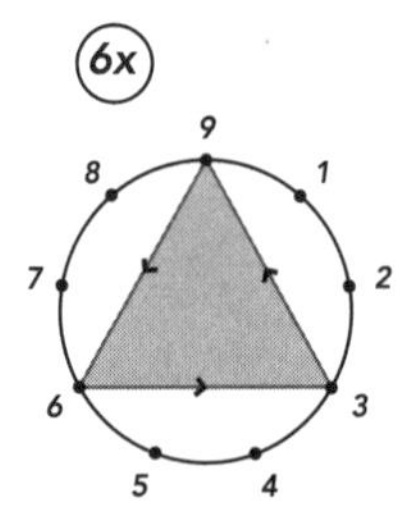
6x
9
8
1
7
2
6
3
5
4

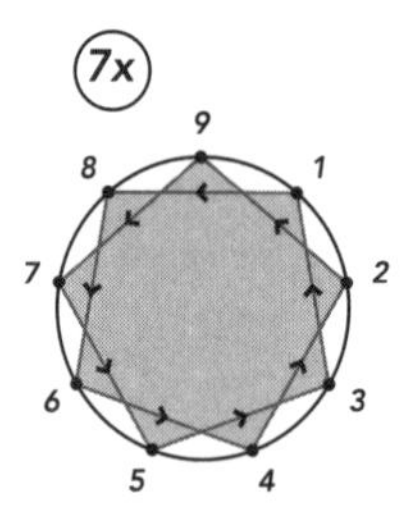
7x
9
8
1
7
2
6
3
5
4

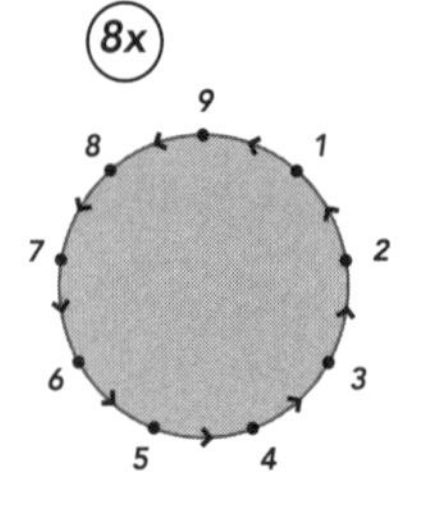
8x
9
8
1
7
2
6
3
5
4

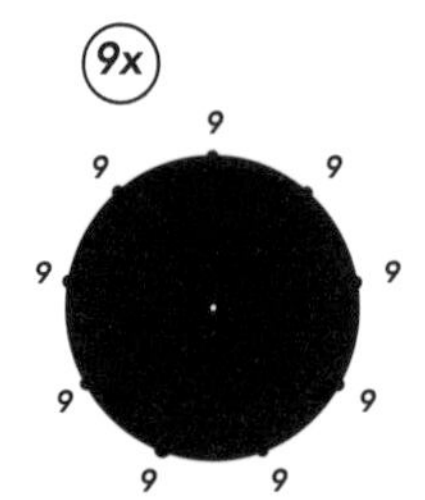
9x
9
9
9
9
9
9
9
9
9

Numbers 1 and 8 rotate in a pure orbit.

Other numbers like 4 and 5 right in the middle of the sequence of 1 – 9 cut across space with high energy star patterns.

And 3 and 6 take up the strong three-sided shape of stability. Their outline, not surprisingly, is the robust triangle.

Multiplication by 2 and 7 form polygons.

The pairings have identical shapes except
that the trace of one is the reverse of the other.

And 9 stays mysterious, unmoving.
It has no orbit except itself.

In all the unseen arithmetic of secret numbers, of addition, subtraction, division and multiplication which revolve around the circle, number nine remains stationary, as if too full of movement to move.

But we are looking at only one, innermost, circle here. There are others.

Mandala

In a magic circle the sigma code first explains all that happens in the secret mirrors of arithmetic. Number nine is seen as a pivot, a point of beginning and ending; it is the place from which all numbers move and circle.

But contemplation of all the multiplication tables and the strange twin pairings of numbers in the sigma circle leads to another and much grander picture.

A Mandala of several circles takes shape and revolves. Like the Mandalas used in meditation, there is an intricacy of pattern and a holding together of an inner harmony between the numbers.

To understand this complete picture we must turn again to the sequences of the multiplication table. We must plot each sequence on radial arms reaching out from the relevant value of the sigma circle, the 2× table starting from point 2, the 3× from point 3, and so on.

These sequences are shown on the opposite page. If these arms are now placed on the sigma circle, in their correct position, we then have a Mandala.

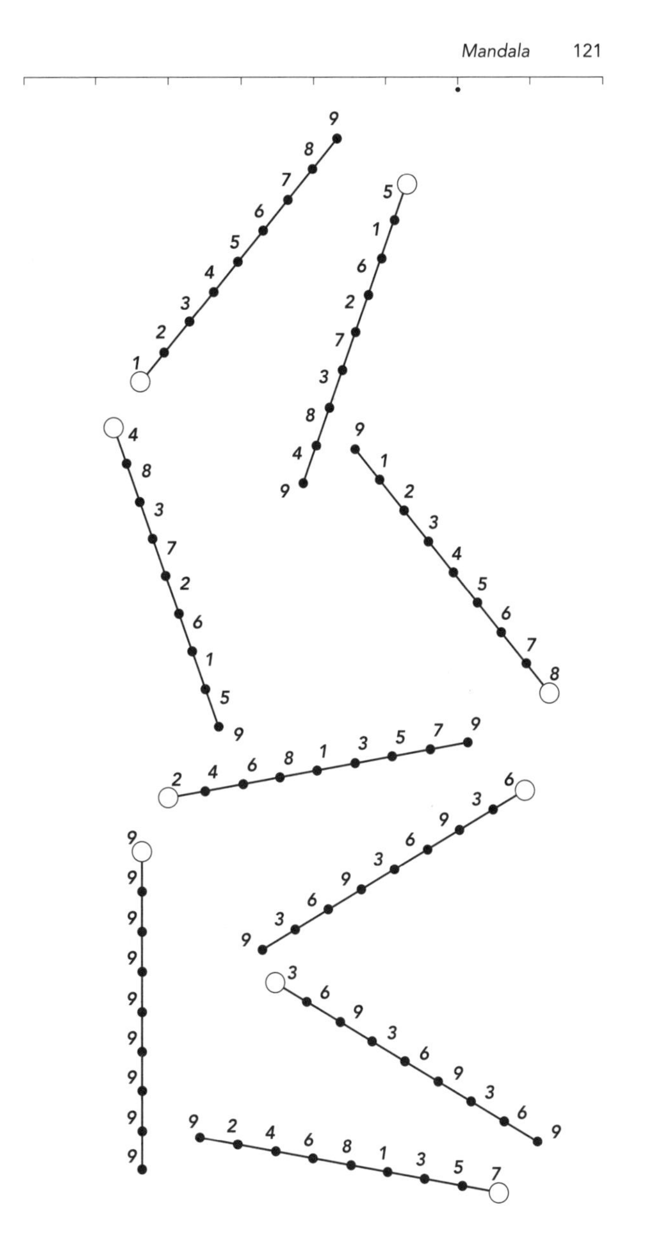

1 2 3 4 5 6 7 8 9
5 1 6 2 7 3 8 4 9
4 8 3 7 2 6 1 5 9
9 1 2 3 4 5 6 7 8
2 4 6 8 1 3 5 7 9
9 3 6 9 3 6 9 3 6
9 9 9 9 9 9 9 9 9
3 6 9 3 6 9 3 6 9
9 2 4 6 8 1 3 5 7

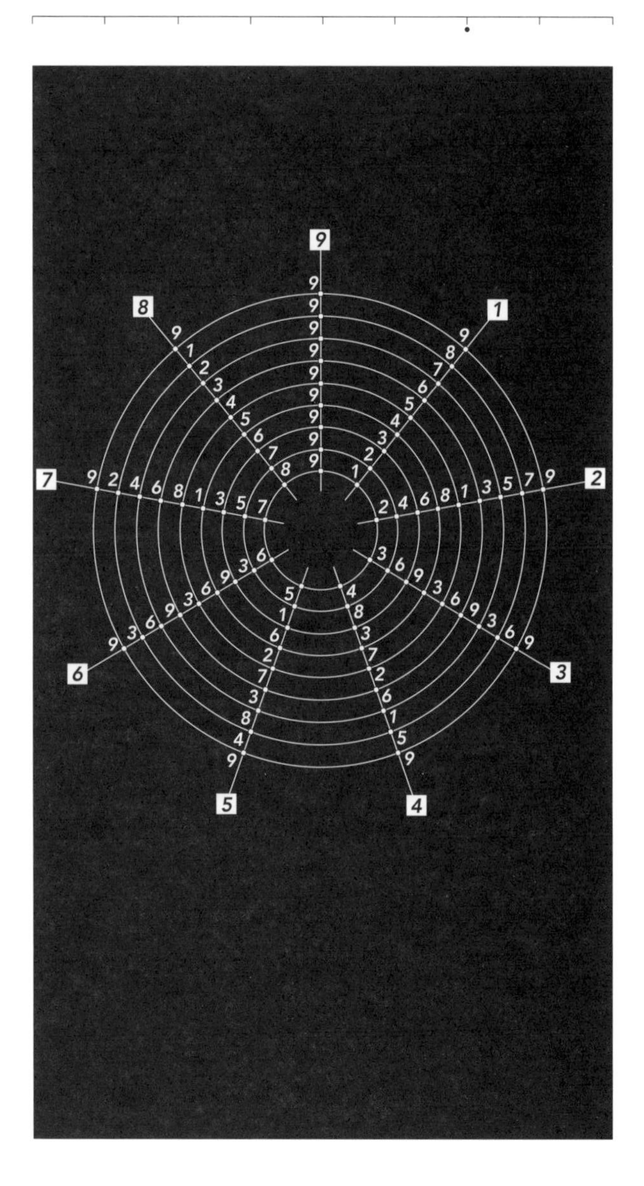
9
9 9 9 9 9 9 9 9 9
1
9 8 7 6 5 4 3 2 1
2
2 4 6 8 1 3 5 7 9
3
3 6 9 3 6 9 3 6 9
4
4 8 3 7 2 6 1 5 9
5
5 1 6 2 7 3 8 4 9
6
6 3 9 6 3 9 6 3 9
7
9 2 4 6 8 1 3 5 7
8
9 1 2 3 4 5 6 7 8

The Complete Revolution

Putting all the parts together reveals nine orbits. It is the blueprint of secret arithmetic. Like the nine planets whirling round the sun, nine orbits circle the centre.

Each ring adds up to nine, as do the numbers along each radiating arm.

The figure is a crystallisation of the sequences of multiplication under the sigma code. The 2× table runs outwards along radiating arm [2], the same sequence also goes round the second circle starting at number 2; the 3× table similarly along arm [3] and around the third circle beginning at number 3, and so on. There is an inner harmony due to this completeness, the straight lines overlaying and combining with the circular motions to produce a strong sense of interlocking and vibration.

Down the mirror plane number 9 dominates, intercepting each orbit; but not only that, the character of nine controls everything. Each circle adds up to nine, each radiating arm adds up to nine as well.

The pattern can be studied for calculations or just looked at, travelling deeper into the mystery behind the numbers. At the outermost orbit number 9 surrounds the figure like a sentinel, guarding and containing everything else in turn.

The outermost layer of nine acts as a boundary and also a mirror.

For example, the radiating arm of [2] has the values – reading from the centre outwards –

2 → 4 6 8 1 3 5 7

Then it hits the outer boundary of nine and travels along the circle coming back on its complement of nine, number 7. The order is now reversed from the inside to the outside along the radiating arm [7]

7 → 5 3 1 8 6 4 2

And so on, the symmetries and the reflections grow.

I lose myself in wonder at the beauty and the intricacy in which the numbers depend on each other, in secret. On the surface one would never imagine this closed circuit of intrigue. In the Mandala of nine, the digits have life-long partnerships, number ONE with EIGHT, TWO with SEVEN, THREE with SIX, FOUR with FIVE. Each pairing has an identical shape – when its multiplication sequence is plotted on the sigma circle, one is the reverse of the other.

Like partners in an eternal dance they turn, their bond being the number nine which holds them together and at the same time keeps them apart. Behind the whole revolution is this special mark, the character that keeps secrets and takes on several disguises – now a boundary, now a mirror, now unseen, and then taking complete hold of another's identity. Whichever way we turn in the circles, we find that all paths out of the labyrinth lead to Σ 9.

It is one grand act of conservation. Each orbit adds up to nine, as does each radiating arm. And in one magnificent homage, all the values of all the circles add up to nine. Everything in this hidden cosmos of numbers comes to the same thing, the uniqueness of 9.

Its component parts, the digits one to eight, hold in tight confidence

this secret
of the first revolution of the

MANDALA.

Hidden by all permutations of arithmetic the numbers of the code spin quietly. In the music of the spheres the sigma code comes together; layer upon layer, orbiting in a secret universe. It is a revolution of numbers beyond that which any of the Elders could imagine. Like the planets in the sky, the code revolves in full nine orbits.

Enjil said numbers are not fixed, but ultimately rotations.

Nine Fixed Points in the Wind

The Elders had never seen it coming.

The young Master was elated as the crowd cheered, each one of them understanding how simple it was. They saw for the first time the basic nature of the numbers themselves, their shapes, the stars and the triangles, the crossing polygons and the spinning discs. Underneath the clutter of the arithmetic was the simple nine number code, which seemed to rotate all around their heads as the young master drew fantastic circles on the blackboard with radiating arms.

But when the boy master had worked out his thesis and drafted those lovely circles of nine, he was not content. He saw the completeness of it but something was not quite right. There was too much balance. It was not satisfying how the nines bordered the figure yet the reflection and the reversals did not catch the spirit of what was in his head.

Enjil wanted a deeper clarity – to match the swirling movements he had first seen – so that the diagram would show better what actually was going on, if that was at all possible.

Like the Elders and wise Masters who had come before him, everyone believed in the perfection of the circle. The world was created round, the sun was round, the stars wrapped around the vast bowl of the sky in a great orbit. Mathematicians knew that numbers had a perfection not corrupted by daily affairs and that the concept of a number was abstract, unchanging. What better shape then, than the perfection of a set of circles?

But to Enjil nine was the number through which everything else flowed. It was also the number that must give reversal and the more he thought of the pattern, of circle upon circle, the more dissatisfied he became. It was too static, too stationary. It did not move. Number nine was full of vigour, it did things to other numbers – it was a catalyst. And yet it had to be stationary to allow other things through it – this was still the problem to solve. But how?

In the afternoon before the day of the examination he saw the breeze gust up and whip up trails of dust in the compound. Minor sand storms that eddied and whirled and zigzagged. He saw spirals and the vortices, as the force in the wind moved his mind into action.

He felt then that number nine was an even greater concept than he had first imagined; it must be like the secret, unseen force of the wind itself, and this would be the force that moved the other numbers. The pattern he sought then must have only one point of reference – number nine itself! Not a circle of numbers, but a condensation of the nature of that first innermost circle. It summed to nine, it had nine parts, it had three hundred and sixty degrees which in turn compressed to nine as three plus six plus zero.

And Enjil wanted the numbers to represent themselves better. He desired that the orbit of eight should be noticeably eight times as large as that of number one, and so on.

He thought the answer must lie in the arrangement of the multiplication tables. It was here that the real secret would be revealed. How did the numbers between one and eight really pair as twins? In intense concentration he meditated on the blueprint of these products, dwelling upon their patterns of multiplications.

The eight times table gave

(8)	(16)	(24)	(32)	(40)	(48)	(56)	(64)	(72)

which in the sigma code was

8	7	6	5	4	3	2	1	9

a reverse of the one times table except for the last figure of nine. If nine itself was a pivot and common to both one and eight, he speculated, then both orbits would connect somehow at nine, the orbit of one number being eight times larger than the other.

Since the sequence of number one is the reverse of number eight, but both have nine in common, then the number nine must be the point that makes the reversal and reflection happen. And this is the case only if the orbit of number one first passes through nine and turns then into the eight times greater orbit of number eight. Where nine occurs there has to be a violent and sudden twist. Otherwise, there would be no reversal.

One orbit has to go one way and then sweep into the other orbit, but going the other way. The point of change has to be a point of reversal, a location of twist.

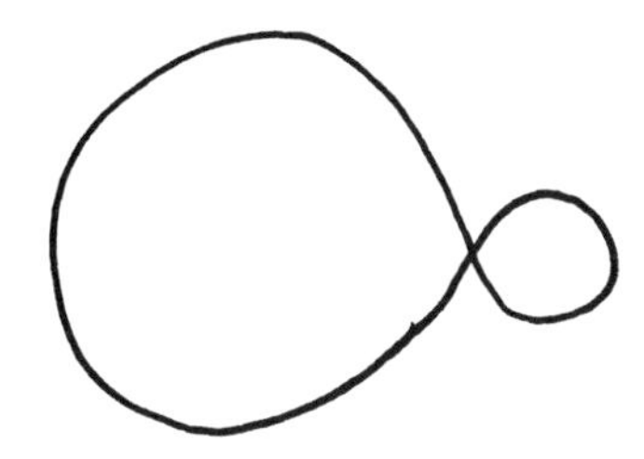

And a strange picture took place in Enjil's head unlike anything he had seen before.

The other pairings of

7 – 2

6 – 3

5 – 4

followed the same logic. One orbit went one way and flowed through nine, in a clockwise sense, reversing itself into its twin partner and then going round in the opposite way, in an anticlockwise orbit.

Quite unlike the stationary circles, energy is released into the numbers so that they spin, one out of the other – a Mandala of a new sort arose before his eyes – the bending and twisting in and out of separate energies, the big and the small, connected by a continuous movement through the eye at the centre of the storm of numbers.

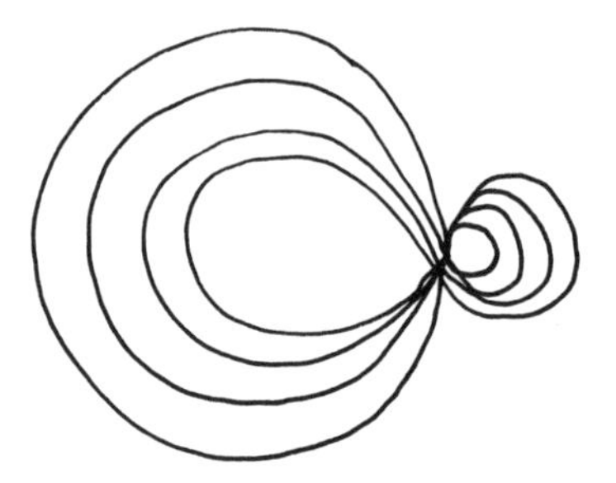

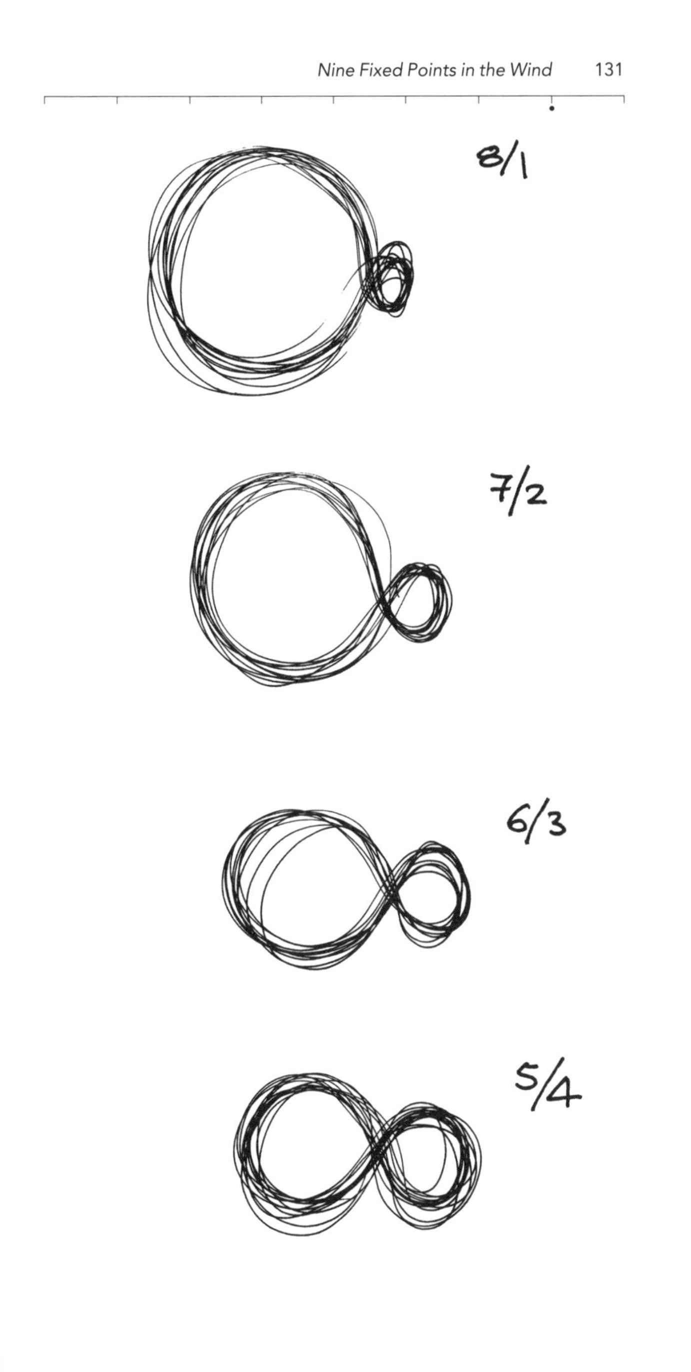
8/1
7/2
6/3
5/4

That eye, the cross-over point, is at nine, the focus of the drawing. And to draw one trace, the focus has to be crossed nine times.

Enjil now had the pattern he wanted, a swirling form that rushed in and out, with vortices spinning. In the paths of reversal, he counted nine crossings. He was joyous; he must indeed have found the fixed points in the wind. The riddle set by the golden woman in the moon was answered.

That night Enjil looked up at the sky to pay homage again to the moon. He laughed and exclaimed to the warm light that bathed down on him:

"You tricked me. There is no fixed point to the
wind, Oh Lady, not one;
and I looked in the wrong
place.

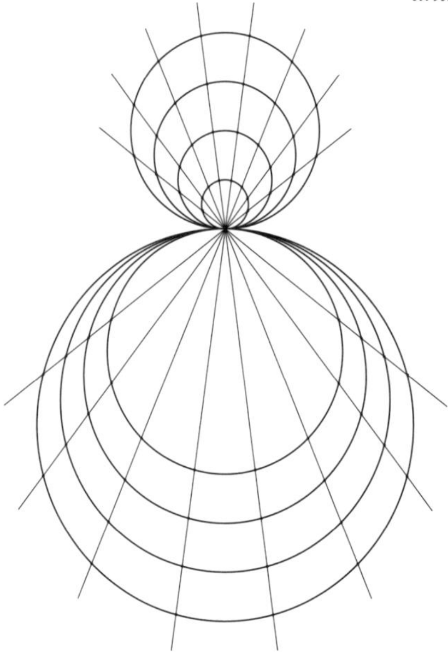

But in looking, I found nine circles and drew them one around the other. Yet all the time there was more. Like the wind I found those circles collapsing and twisting and passing through each other in different orbits, in different ways. Indeed, the wind has not just one fixed point, Oh Soma, but nine. Did you know that? I now laugh and play with numbers and my debt is to you, Soma – the fair one who makes light out of new beginnings."

He unfurled the large drawing he had done in different coloured inks and held it up to the moon.

He said, "Just for tonight my lady your secret is mine. Tomorrow it will be with the Elders and the keen eyes of the crowd. After that I pray the orbits move into the worlds of people's minds and playfully keep turning and twisting, bringing out the free spirit which hides in each one of them."

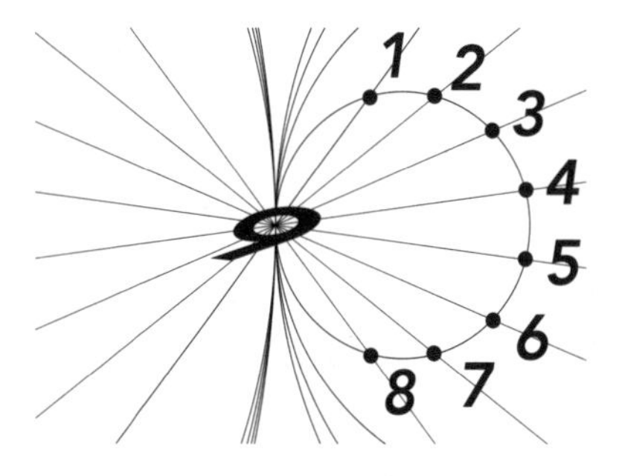

Enjil smiled then at his wonderful secret, at the number of worlds of nine he had found and bid the lady in the moon goodnight.

For want of a better name let us call this new Mandala, the Strange One.

All the symmetries of the original Mandala are to be found in the new drawing if one studies it.

The radial lines intersecting the orbits have values that are reflected through the eye of number 9. On the left hand side of the diagram, looking at it sideways, the values on each radiating line add up, anticlockwise, from $\Sigma 8 - \Sigma 1$. On the right hand side, we have the opposite pattern from $\Sigma 1 - \Sigma 8$, clockwise. This is a clearer picture on the nature of reversals inherent in the sequences of the numbers than in the first circular Mandala.

> *The ninth number in the sequence is the hub of the diagram, number nine itself IS the magic eye in the vortex of spinning numbers.*

The Strange One catches this essence perfectly. In the dynamic that swirls around, number nine is stationary, serene and unmoving.

The Vedic Drawing of the Elders held all multiplication in a square where nine was a border and a centre to the diagram. In the first Mandala Enjil discovered, nine became the outermost circle and a mirror plane through which all the other orbits are reflected. But in the Strange Mandala, nine is at last given its true power, of centre and border, in spectacular patterns of shifting and twisting orbits. To Enjil it was like seeing great firewheels in the sky, exploding and turning, first one way and then burning in a flash to go the other way. And through the curling eye of sparks that was number 9, all the other trajectories of numbers moved.

Enjil knew that the Elders and the village folk were used to seeing symmetry in their constructions. Did not the heavens move in celestial orbits, in great circles around the centre of the Earth, was not the sun round and perfect? Were not the prayer wheels and the Mandalas of meditation equal in part about centre lines? But this is to see symmetry as the balance of separate parts which means an assumption that each part is distinct. What if everything connected, one to the other? Then true symmetry would be the centre that contains everything or through which everything flows. In this sense the strange Mandala to Enjil was perfect, the point 9 being the centre of symmetry. If we could really understand how the heavenly bodies moved we would find that one orbit influenced the other, and that the real point of symmetry would not be the centre of the sun but some other place, deep in the blackness, which would take into account the stars and that burning point of our sun.

Somewhere in this great void would be the other Strange Mandala, even greater than the one Enjil had drawn and dreamt of, and even more marvellous with its twistings and turnings, passing through the shape of a special number, one that would hold the character to all other movements. And Enjil knew that this special number would be at the heart of a great spiral, the beginning point of spin and the spur of all movements.

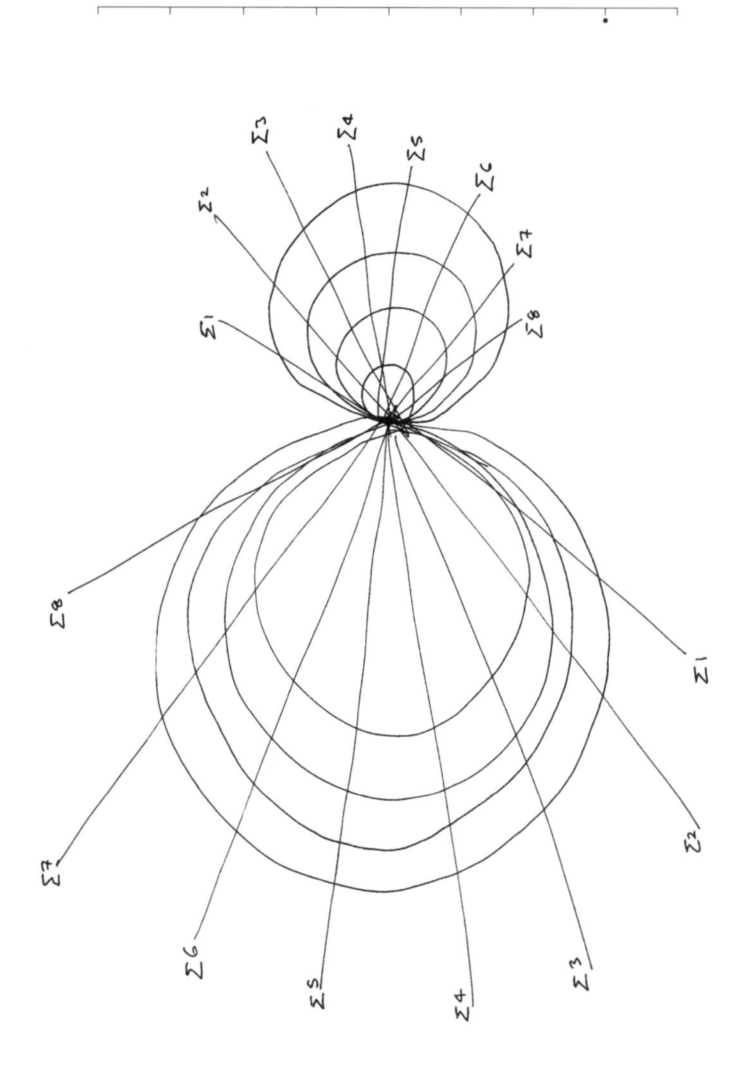
Σ1
Σ2
Σ3
Σ4
Σ5
Σ6
Σ7
Σ8
Σ8
Σ7
Σ6
Σ5
Σ4
Σ3
Σ2
Σ1

When Enjil drew the first Mandala of perfect circles he then followed it with the second Mandala, the Strange One. He explained that this second set of orbits was but the same as the first set of circles, except that they were shown in a better way to express the real character of nine in relation to the other numbers, one to eight.

But though Enjil loved his Strange Mandala and felt comfortable with the shifting orbits, he knew it was an extraordinary jump for those who wanted identical symmetry and patterns of equal parts. So Enjil gave them one last demonstration to set them talking. To end his thesis, copying from a parchment pinned to the board, the young Master drew quickly the third and last Mandala, the one of Powers and gave the Elders and the gathered crowd their last astonishment.

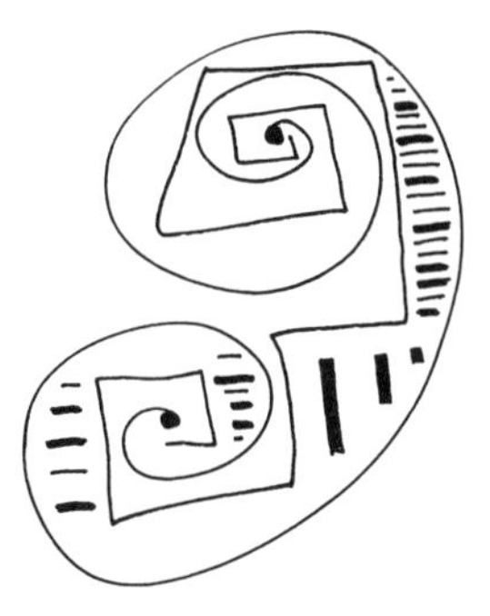

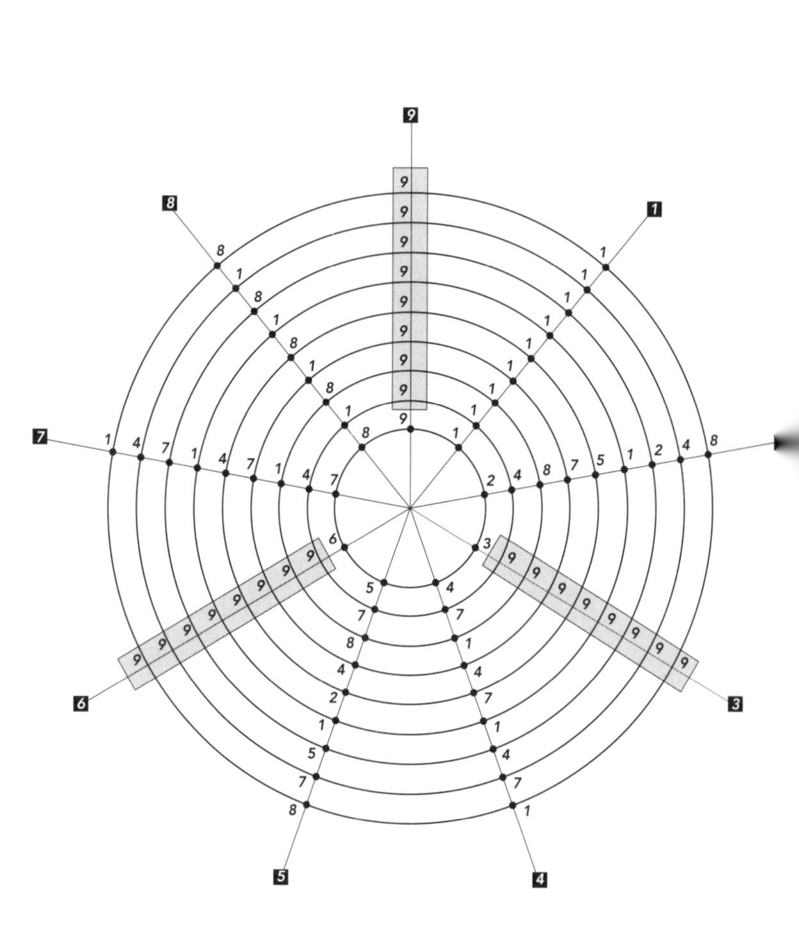
9
8
1
7
6
3
5
4

Powers

The Elders could just about bring themselves to admit that addition and subtraction fitted neatly within the logic of a set of circles, derived from adding up the integers of numbers by applying the sigma code. But powers? Those exponentially jumping values of doubling, tripling, quadrupling.... Could that escalation of numbers have set patterns, even though the numbers grew so rapidly beyond simple calculation?

Enjil had gone too far, they whispered amongst themselves as the boy gave them his crooked smile and picked up the chalk again. To their mind his voice was cheeky, arrogant, and insolent, announcing that the power expansion of numbers unfolded in set ways and was not open ended as they may have thought. The sun, shining on the young Master, threw sharp shadows into the holes of his ravaged pockmarked face. The black spots made him look sinister. Were they in the hands of a demon, they asked, as some of them feared?

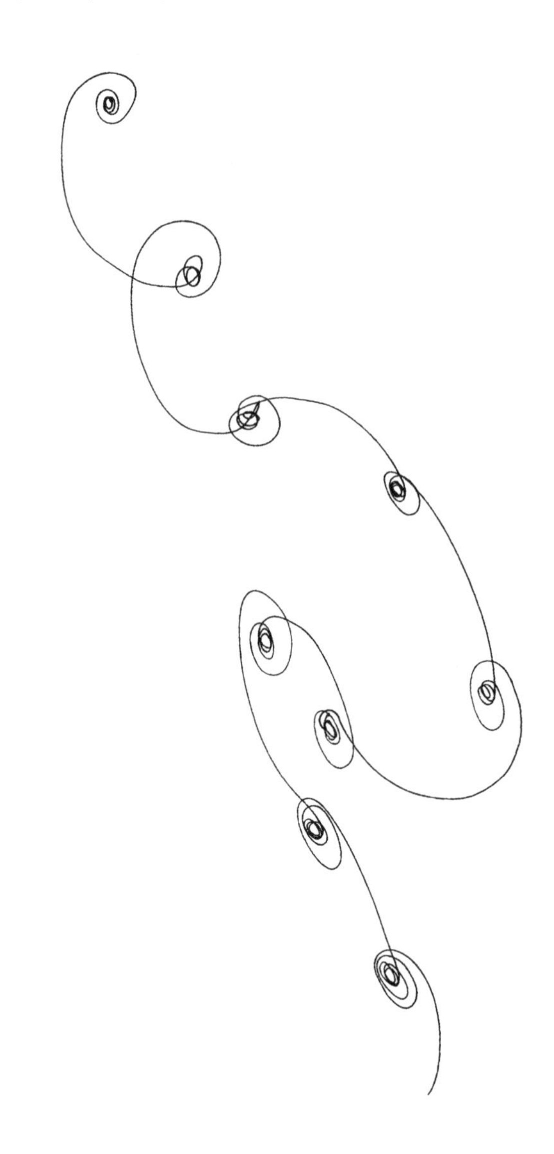

Then Enjil began his demonstration by writing out the numbers, adding the digits to reduce the powers to one number only, and revealing the patterns to the astonished Elders. The adoring crowd who by now had accepted him as the light of the Academy settled down and went quiet, awaiting the demonstration as when a new comet comes into a black sky.

Though this time the concept of the powers of numbers was beyond their daily arithmetic, the way Enjil reduced the problem to simple numbers made the crowd believe they had been shown a great mystery, that unto now had been hidden; and even though they may never use doubling and quadrupling in their everyday affairs, the fact that even such big jumps were controlled in set ways, underneath the surface of numbers, was something they found extraordinary.

The world and the huge jumps to the stars were perhaps then not so great and unbelievable – perhaps even invisible ways were all around, if only they knew how to look. So they willed the young Master success. They watched him limp from one end of the blackboard to the other as he wrote out his belief in the expansion of numbers in an elegant way. And they began clapping when they saw the fun of it all.

Enjil wrote out his workings in a neat hand.

The workings

The sequence for N^2 where N = 1, 2, 3 ... is

1^2	2^2	3^2	4^2	5^2	6^2	7^2	8^2	9^2 ...
1	4	9	16	25	36	49	64	81 ...

If the integers of the numbers in the second row are added and reduced to one number, then we have

1	4	9	7	7	9	4	1	9 ...

As the expansion continues from 10^2 to 18^2 we have

10^2	11^2	12^2	13^2	14^2	15^2	16^2	17^2	18^2
100	121	144	169	196	225	256	289	324

The same rule is applied to reduce these numbers to one value

1	4	**9**	7	7	**9**	4	1	**9**

In fact, as the square powers are plotted further, the code numbers keep repeating. Every third, sixth and ninth value in the sequence is 9.

Similarly the expansion N^3 for N =1 to 9 gives

1	*8*	*27*	*64*	*125*	*216*	*343*	*512*	*729*

which reduces, according to the sigma code as

1	*8*	***9***	*1*	*8*	***9***	*1*	*8*	***9***

Here the rhythm is 1, 8, 9 with the third, sixth and ninth values acting as anchor points, just as in the sequence for N^2.

Similarly the expansion of N^4 is

1	*16*	*81*	*256*	*625*	*1296*	*2401*	*4096*	*6561*
1	*7*	***9***	*4*	*4*	***9***	*7*	*1*	***9***

Again the pattern emerges here of every third value being 9, and the sequence being nine numbers, beginning with one and ending with nine.

If the calculations are continued to higher values of N and the code sequences are written out from N^1 to N^9 we arrive at:

N^1	1	2	3	4	5	6	7	8	**9**
N^2	1	4	**9**	7	7	**9**	4	1	**9**
N^3	1	8	**9**	1	8	**9**	1	8	**9**
N^4	1	7	**9**	4	4	**9**	7	1	**9**
N^5	1	5	**9**	7	2	**9**	4	8	**9**
N^6	1	1	**9**	1	1	**9**	1	1	**9**
N^7	1	2	**9**	4	5	**9**	7	8	**9**
N^8	1	4	**9**	7	7	**9**	4	1	**9**
N^9	1	8	**9**	1	8	**9**	1	8	**9**

These sequences can help find other relationships between the powers of numbers. Because the arithmetic is reduced to single values, it is easy to test various speculations about how numbers relate to each other. I found interesting patterns very quickly using the code values rather than carrying out the actual multiplication. All one has to do is obtain the sigma value, the rest is easy.

The interesting feature is that each of the sequences of nine values, when added up horizontally, gives the rhythm 9 – 6 – 9 – 6 – 9 .... That is the row for N^1, in the above table adds up to 9, the row for N^2 adds up to 6. The row for N^3 adds up to 9 and so on.

If each sequence of the power expansions is written out in a circle of nine values, we obtain a picture of concentric orbits say for N^2 and N^3, the following is then true:

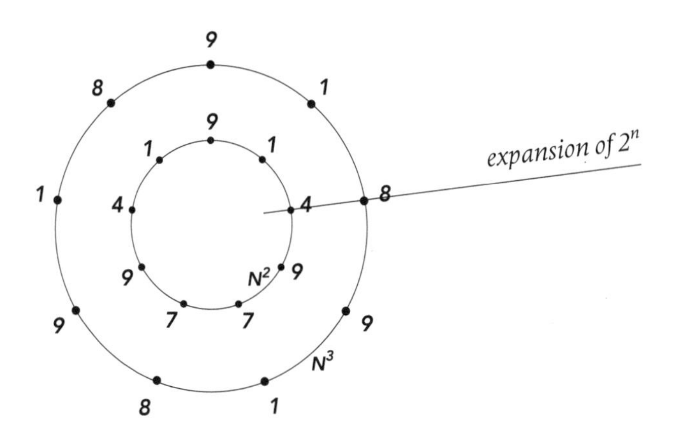

The inner circle is the sequence for N^2. This means the whole expansion of N^2, sits along this inner circle with values 1 – 4 – 9 – 7 – 7 – 9 – 4 – 1 – 9 spinning round the circumference.

And the outer circle is the sequence for the expansion of N^3, with the values 1 – 8 – 9 repeating. Because the sequences have only nine values they fit around each other neatly, just as in the first Mandala. There is also a flow outwards along radiating arms of the powers of a given number. The line for 2^n is shown, i.e. $2^2 = 4$, $2^3 = 8$, etc.

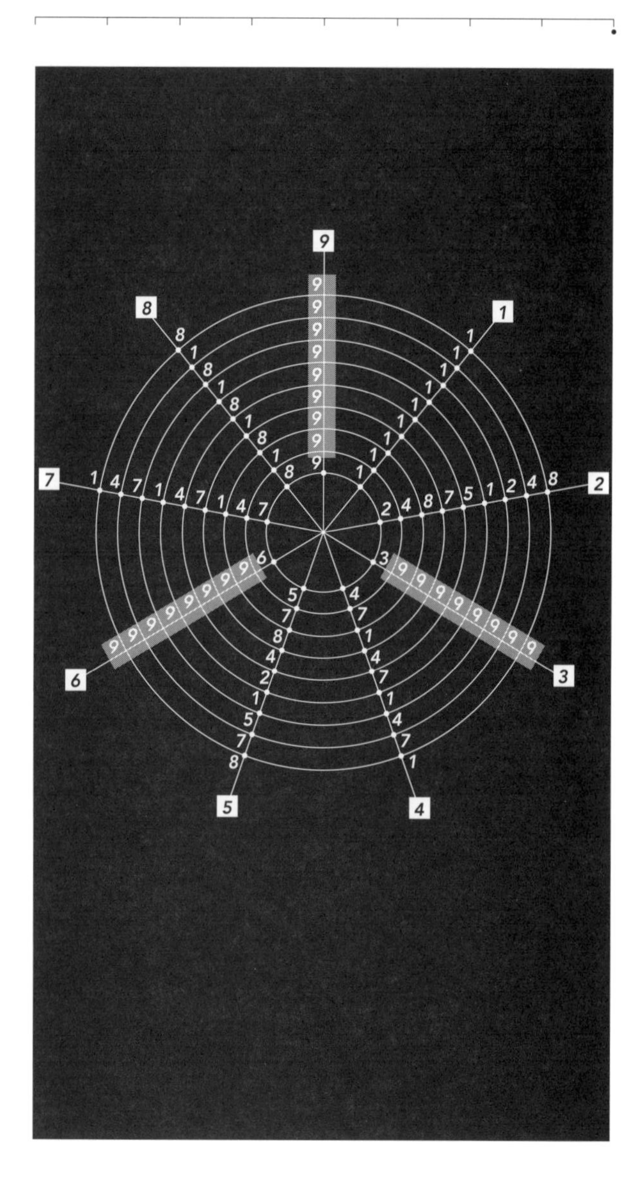
9
9 9 9 9 9 9 9 9 9
1
1 1 1 1 1 1 1 1 1
2
2 4 8 7 5 1 2 4 8
3
3 9 9 9 9 9 9 9 9
4
4 7 1 4 7 1 4 7 1
5
5 7 8 4 2 1 5 7 8
6
6 9 9 9 9 9 9 9 9
7
7 4 1 7 4 1 7 4 1
8
8 1 8 1 8 1 8 1 8

Power Mandala

The innermost circle is the expansion of N^1, the sigma code itself, the numbers 1 – 9.

The second circle relates to N^2. The third circle relates to N^3 and so on, where N = 1, 2, 3, ...

But each radiating arm is the sigma sequence of a fixed number being raised successively to higher powers.

For example, the arm starting at number 2 in the innermost circle shows the sequence

2 4 8 7 5 1 2 4 8

which is the sigma value of the power expansion of

$2^1 \quad 2^2 \quad 2^3 \quad 2^4 \quad 2^5 \quad 2^6 \quad 2^7 \quad 2^8 \quad 2^9$

Each radiating arm sums up to a multiple of nine and successive orbits have the sigma pattern 9 – 6 – 9 – 6 – 9 – 6 . . . when the values of each orbit are added up, starting from the innermost circle.

The strong radiating lines of 9 divide the Mandala into three sections.

It can also be seen that along each radiating arm some patterns are short, before they repeat, like that for number 4. Numbers 2 and 5 have longer sequences before they are repeated. But the numbers 3, 6 and 9 raised to successive powers are held by the fixed value of nine.

The patterns that repeat are:

1^N	–	*1 …*
2^N	–	*4 8 7 5 1 2 …*
3^N	–	*9 …*
4^N	–	*7 1 4 …*
5^N	–	*7 8 4 2 1 5 …*
6^N	–	*9 …*
7^N	–	*4 1 7 …*
8^N	–	*1 8 …*
9^N	–	*9 …*

The power circles show that though a number seemingly jumps rapidly as it is raised to higher powers, underneath the surface there is a set of repeating cycles.

There is conservation at work, in secret.

So we go back in time to that first circle of the compound in the Academy, with the sun shining down on the blackboard and on Enjil as he draws his power circles. Not only have additions, subtractions, multiplications and divisions been held in the power of just nine orbits, now he demonstrates how all the possibilities of expanding numbers in squares, cubes, and quadrupling and so on are also held around the circumference and radius of just nine circles, one within the other.

As Enjil adds the numbers within each sector, along each curve, and outwards along each radius, he shows them how nine and its stepping stones of three and six run through the whole construction of this newest Mandala. The mathematics of the powers of numbers are beyond the humble village folk who keep arithmetic simple, to bartering for grain and livestock, but they still smile and cheer for they understand how complicated things can be broken down and made simple. The Elders are different; they are clever and know all about numbers – or so they thought until this fateful day of the Examination, which turns into an examination of themselves. Now they have to look again at how they see the dust of arithmetic after what Enjil has shown them, the reflections that turn in the precious mirrors, visions, covered with numbers in secret ....

... Enjil showed them that numbers were more than just counters in calculations. They had a secret life. If anything, these hidden reflections were more exciting than the actual images of arithmetic. For on the surface every calculation was a forced answer with one outcome only; and the possible outcomes were a million times a million and more in all the calculations one could do. But it was not just the outer event that mattered, that was nothing if an inner life could not be guessed at.

In this way Enjil argued for the jewels he had found, hidden in the dust of calculations. As he said, if calculations and all the sums we do are just patterns of an untamed wind, blowing any which way it pleases, what cannot change is the character that seeds the wind, the fixed points which structure and raise shape and whose force always gives direction.

The powers of numbers jumped like those ghosts in a storm, growing larger and more frightening every time. But in the leaps, a rhythm, a pulse that no one could see turned, in repeating ways, never changing no matter how big the jumps. No one need be afraid, he said.

Enjil's proposal was that his Mandalas were fixed orbits which gave the essence of all the movements of numbers. In secret mirrors, they held the inner reflections.

Like with people, numbers had a hidden nature. And even though we may do so much in the outward appearance of our lives, what mattered was the inside shape of a person, he argued, that gave character to everything else.

Even though people were all different in various circles of power and influence, underneath it all they shared a set of values, love of family and fear of God. People everywhere held to some sort of secret code. Even if they did not talk about it their actions were governed by an inner faith. In the same way Enjil said numbers were blown this way and that by the fate of calculation and worldly actions, all appearing to be different; but underneath it all a hidden mantra turned, holding sway in a myriad world of change.

He said the Mandalas of nine were but the fixed points in a set of beautiful meditations, on the meaning of numbers.

Movements of Nine

Part Two

The Search . . .

The intrigue of nine does not stop with the Mandalas. The movements of nine go on.

In a way, these movements are the heart of the book. Enjil and the Mandalas of nine were but the catalyst and inspiration; now comes the unfolding, the discoveries of what lies buried in higher arithmetic and mathematical topographies.

I said at the outset that nothing original stops, it is just taken over. As one story ends another begins. The Mandalas are now mine and Enjil has become my talisman; a charm that works its spell under the different disguises of arithmetic and mathematics, for if I look with the sigma code I find things. I see hidden paths. Like ghosts, footprints emerge. What is complicated or plain bewildering suddenly becomes simple.

By looking below the surface we find surprises.

For those who search out more from the numbers the trail now goes on to follow the movements of nine, into more intriguing and deeper constructions; we enter a world where the hypnotic trace and secret mantras continue in remarkable ways.

The reader is encouraged to journey through, marking trails, so that in time his or her own journeys may be made into the force of nine.

And somewhere inside me Enjil, with a great big smile on his face, is urging us to look further. The quest, as he puts it, is for the crucible of numbers, for the fixed points in the wind that do not move, though everything else around them is changing all the time.

What follows is an open invitation, to the reader, to join the sigma circle.

Cycles and Patterns

Circular Numbers

A number well known to number theorists that seems to cycle itself upon multiplication is 142 857. When multiplied by 1, 2, 3, etc. up to 6, the answer is the same number but moved along a few places.

$1 \times 142\,857 = 142\,857$
$2 \times 142\,857 = 285\,714$
$3 \times 142\,857 = 428\,571$
$4 \times 142\,857 = 571\,428$
$5 \times 142\,857 = 714\,285$
$6 \times 142\,857 = 857\,142$

The number has a circular nature.

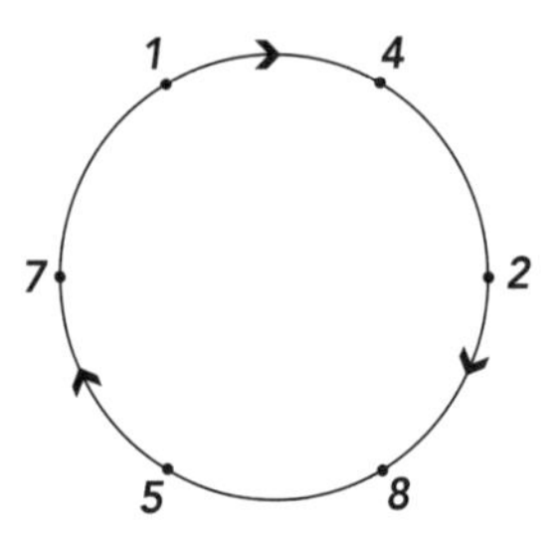

But what I cannot find in the number theory books is how this number relates to nine and our invention of the sigma circle.

And here is a surprise!

$\Sigma\, 142857 = 9$

We know then that every multiplication of this number, by one or two or whatever, will never change the sigma value because as we have seen when 9 enters multiplication, the product always adds up to nine.

When this number is plotted on the sigma circle it shows a symmetry about the vertical axis, about the lone and proud figure of 9.

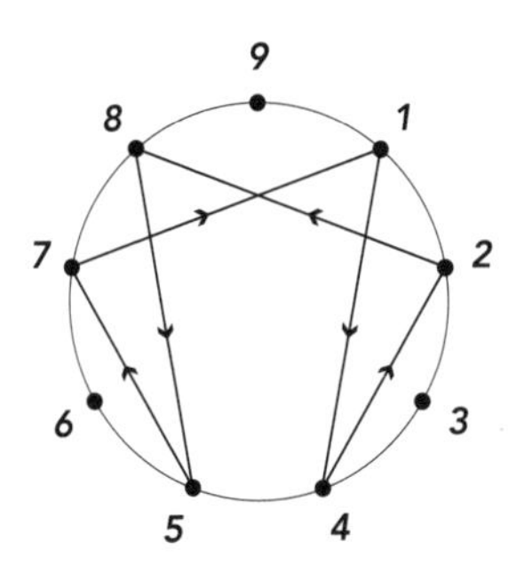

Cyclic numbers are exceedingly rare and are hard to find. Can the sigma circle help to find others?

The number 0 1 2 3 4 5 6 7 9 has cyclic properties upon multiplication as shown below:

012345679	×	*19*	=	*234567901*
	×	*28*	=	*345679012*
	×	*37*	=	*456790123*
	×	*46*	=	*567901234*
	×	*55*	=	*679012345*
	×	*64*	=	*790123456*
	×	*73*	=	*901234567*

Remember 1089?

As a fraction, the number that leads to tricks with reversals has an inversion of the digits 1 – 9.

1/1089 = 0.00091827364554637281 9 . . .

The pattern on page 169 gives a picture of this sequence reading from right to left, starting at the bottom, 9 – 1   8 – 2   7 – 3   6 – 4   . . .

The number on the left comes from the fraction $^1/_{81}$ which has the repeating cycle to its expansion of .012345679. If we forget the decimal and just take the numbers that repeat then we have this number. When it is multiplied by the 9× table we get an interesting sequence in a nine digit pattern:

---

1 2 3 4 5 6 7 9 × 09 = 1 1 1 1 1 1 1 1 1

1 2 3 4 5 6 7 9 × 18 = 2 2 2 2 2 2 2 2 2

1 2 3 4 5 6 7 9 × 27 = 3 3 3 3 3 3 3 3 3

1 2 3 4 5 6 7 9 × 36 = 4 4 4 4 4 4 4 4 4

1 2 3 4 5 6 7 9 × 45 = 5 5 5 5 5 5 5 5 5

1 2 3 4 5 6 7 9 × 54 = 6 6 6 6 6 6 6 6 6

1 2 3 4 5 6 7 9 × 63 = 7 7 7 7 7 7 7 7 7

1 2 3 4 5 6 7 9 × 72 = 8 8 8 8 8 8 8 8 8

1 2 3 4 5 6 7 9 × 81 = 9 9 9 9 9 9 9 9 9

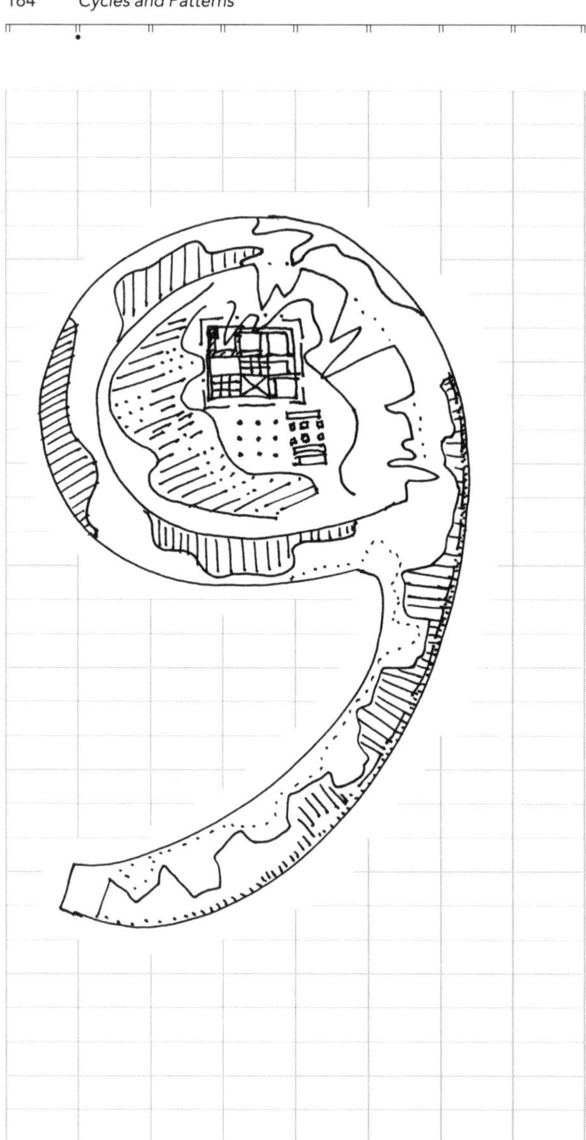

Patterns

The essence of mathematics is to look for patterns.

Our minds seem to be organised to search for relationships and sequences. We look for hidden orders.

These intuitions seem to be more important than the facts themselves, for there is always the thrill at finding something, a pattern, it is a discovery – what was unknown is now revealed. Imagine looking up at the stars and finding the zodiac!

Searching out patterns
is a pure delight.

Suddenly the counters fall into place and a connection is found, not necessarily a geometric one, but a relationship between numbers, pictures of the mind, that were not obvious before. There is that excitement of finding order in something that was otherwise hidden.

And there is the knowledge that a huge unseen world lurks behind the façades we see of the numbers themselves.

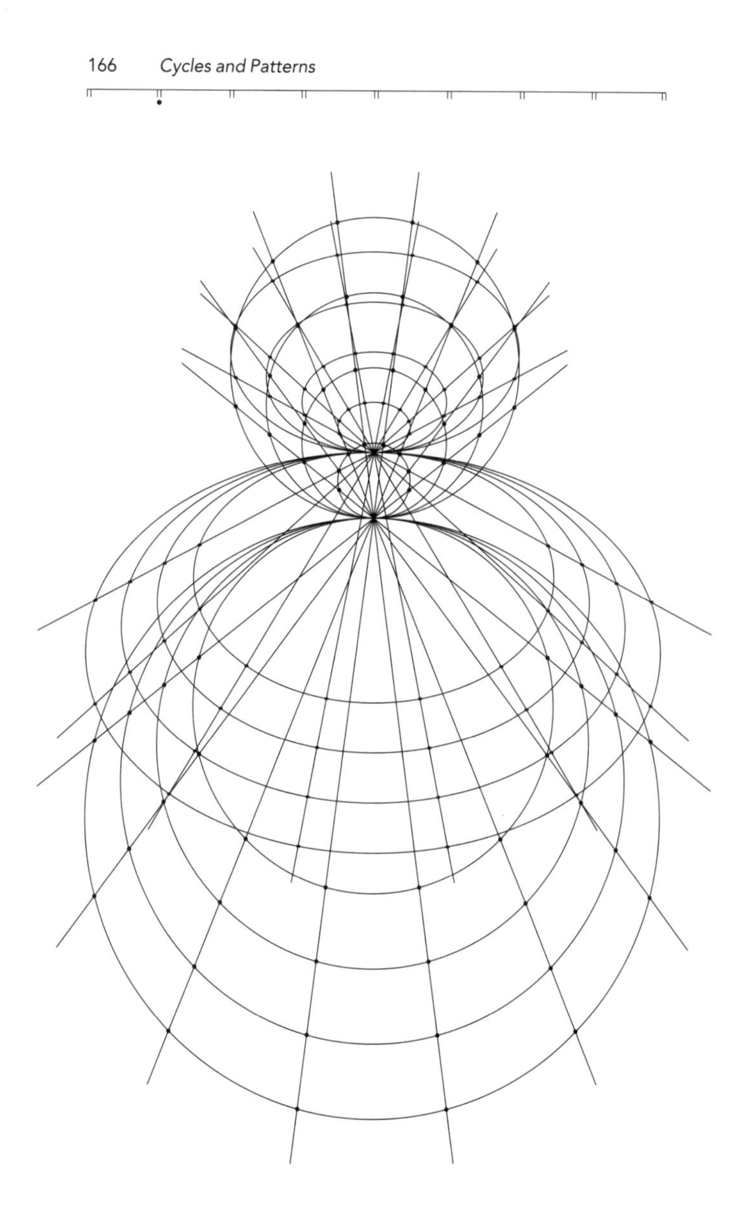

Patterns with all nine digits

$$
\begin{aligned}
0 \times 9 &+ 1 = 1\\
1 \times 9 &+ 2 = 11\\
12 \times 9 &+ 3 = 111\\
123 \times 9 &+ 4 = 1111\\
1234 \times 9 &+ 5 = 11111\\
12345 \times 9 &+ 6 = 111111\\
123456 \times 9 &+ 7 = 1111111\\
1234567 \times 9 &+ 8 = 11111111\\
12345678 \times 9 &+ 9 = 111111111
\end{aligned}
$$

$$
\begin{aligned}
0 \times 9 &+ 8 = 8\\
9 \times 9 &+ 7 = 88\\
98 \times 9 &+ 6 = 888\\
987 \times 9 &+ 5 = 8888\\
9876 \times 9 &+ 4 = 88888\\
98765 \times 9 &+ 3 = 888888\\
987654 \times 9 &+ 2 = 8888888\\
9876543 \times 9 &+ 1 = 88888888\\
98765432 \times 9 &+ 0 = 888888888
\end{aligned}
$$

$$
\begin{aligned}
1 \times 8 &+ 1 = 9\\
12 \times 8 &+ 2 = 98\\
123 \times 8 &+ 3 = 987\\
1234 \times 8 &+ 4 = 9876\\
12345 \times 8 &+ 5 = 98765\\
123456 \times 8 &+ 6 = 987654\\
1234567 \times 8 &+ 7 = 9876543\\
12345678 \times 8 &+ 8 = 98765432\\
123456789 \times 8 &+ 9 = 987654321
\end{aligned}
$$

1/099	*.010101010101010101010 . . .*
1/198	*.005050505050505050505 . . .*
1/297	*.003367003367003367003 . . .*
1/396	*.002525252525252525252 . . .*
1/495	*.002020202020202020202 . . .*
1/594	*.001683501683501683501 . . .*
1/693	*.001443001443001443001 . . .*
1/792	*.001262626262626262626 . . .*
1/891	*.001122334455667789001 . . .*
1/990	*.001010101010101010101 . . .*

The above fractions have two features to their denominators. Each is Σ 9 and to each side of the fixed middle figure there is a rise and fall of the digits 1 – 9.

When these fractions are added up, the result has an intriguing secret.

0.02958553791887125220 2

Correct to nine decimal places this is

0.029585538

Remarkably, this adds up to 9.

$$19 + 28 + 37 + 46 + 55 + 64 + 73 + 82 + 91 = 495 = \Sigma 9$$

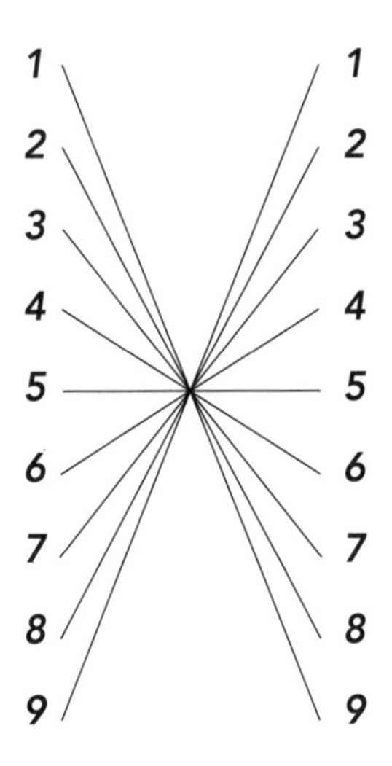

$$19^2 + 28^2 + 37^2 + 46^2 + 55^2 + 64^2 + 73^2 + 82^2 + 91^2 = 32\,085$$

The individual numbers above raised to the same power give a sequence whose sum equals $\Sigma\, 9$:

$$\Sigma\,(19^n + 28^n + 37^n + 46^n + 55^n + 64^n + 73^n + 82^n + 91^n) = \Sigma\, 9$$

where n = 0, 1, 2, 3, . . . etc.

The ③–⑥–⑨ Triangle in the Sigma Circle

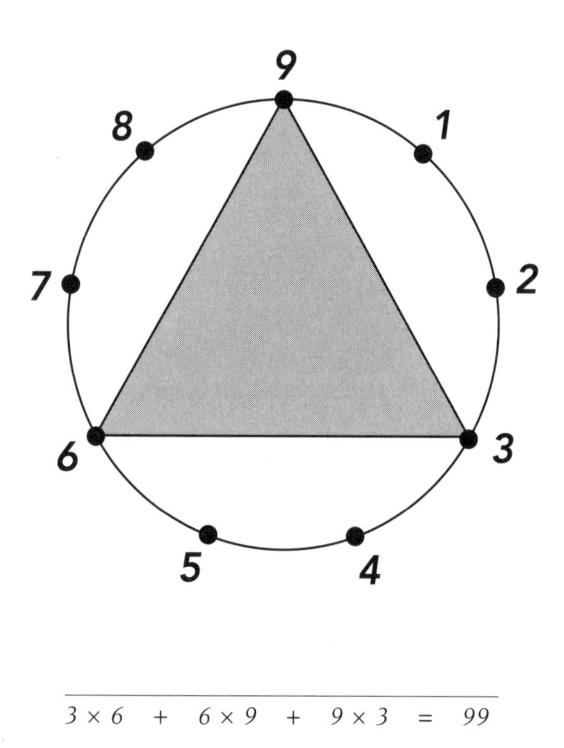

$3 \times 6 \quad + \quad 6 \times 9 \quad + \quad 9 \times 3 \quad = \quad 99$

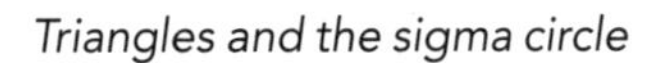

Triangles and the sigma circle

Base 8 – 1

$8 \times 7 \times 1 =$	56
$8 \times 6 \times 1 =$	48
$8 \times 5 \times 1 =$	40
$8 \times 4 \times 1 =$	32
$8 \times 3 \times 1 =$	24
$8 \times 2 \times 1 =$	16
$8 \times 9 \times 1 =$	72
	288

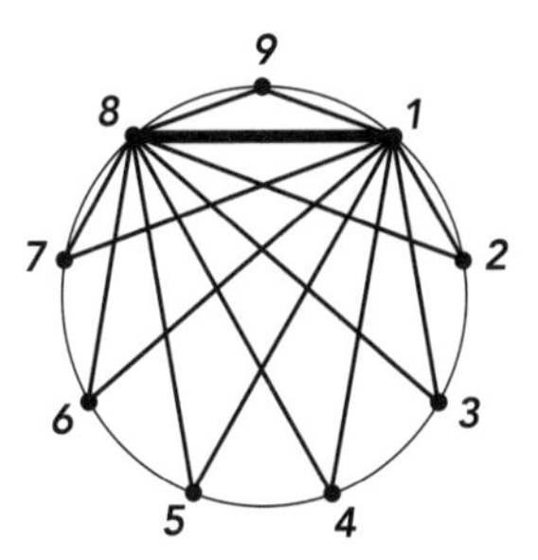

Base 7 – 2

$7 \times 6 \times 2 =$	84
$7 \times 5 \times 2 =$	70
$7 \times 4 \times 2 =$	56
$7 \times 3 \times 2 =$	42
$7 \times 1 \times 2 =$	14
$7 \times 9 \times 2 =$	126
$7 \times 8 \times 2 =$	112
	504

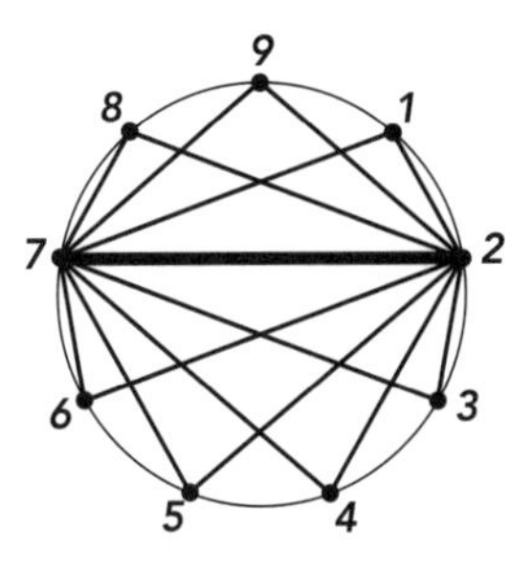

Base 5 – 4

$5 \times 3 \times 4 =$	60
$5 \times 2 \times 4 =$	40
$5 \times 1 \times 4 =$	20
$5 \times 9 \times 4 =$	180
$5 \times 8 \times 4 =$	160
$5 \times 7 \times 4 =$	140
$5 \times 6 \times 4 =$	120
	720

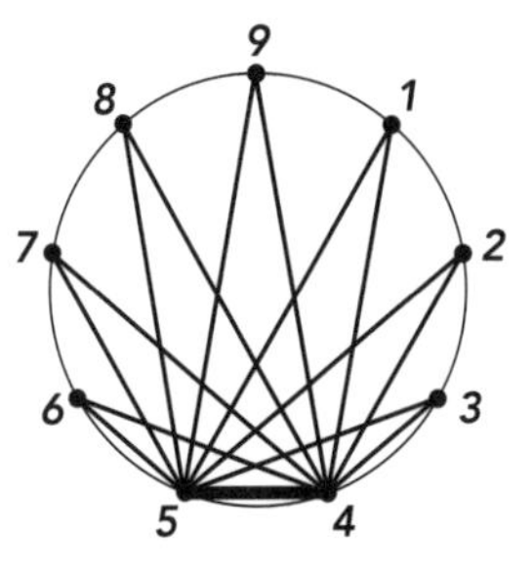

A horizontal base to the triangles produces answers that add up to 9. Inclined bases give different answers. Try it!

Patterns across the sigma circle

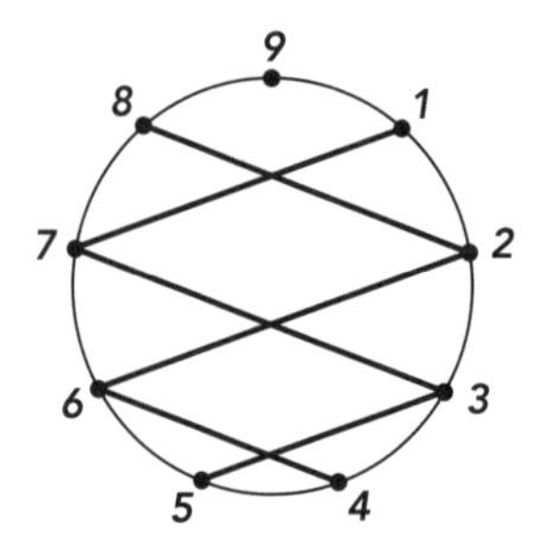

$28 + 71 = 99$
$37 + 62 = 99$
$46 + 53 = 99$
$55 + 44 = 99$
$64 + 35 = 99$
$73 + 26 = 99$
$82 + 17 = 99$

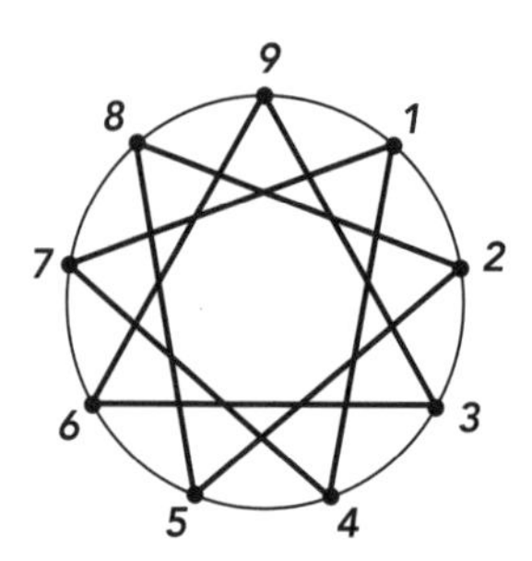

This is the same calculation as multiplying out the numbers on the keys of a calculator pad, vertically.

$$1 \times 4 \times 7 \quad + \quad 2 \times 5 \times 8 \quad + \quad 3 \times 6 \times 9 \quad = \quad 270$$

Across the sigma circle

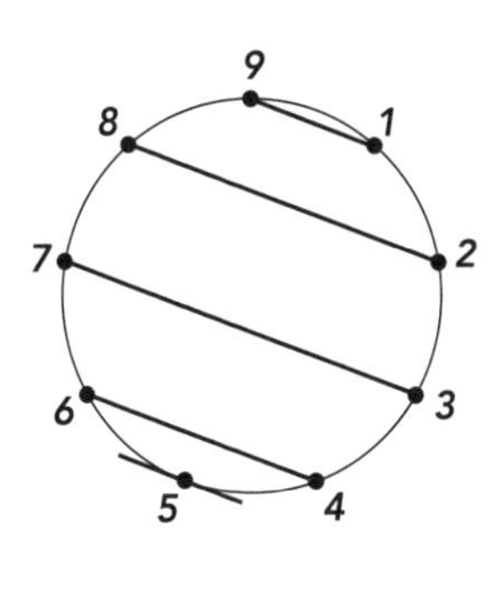

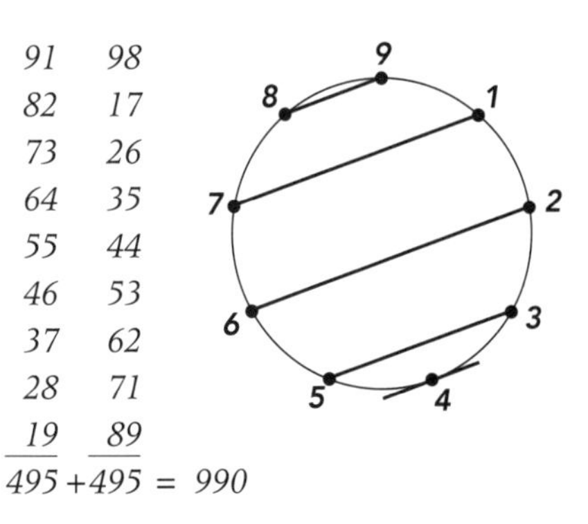

91	*98*
82	*17*
73	*26*
64	*35*
55	*44*
46	*53*
37	*62*
28	*71*
19	*89*

495 + 495 = 990

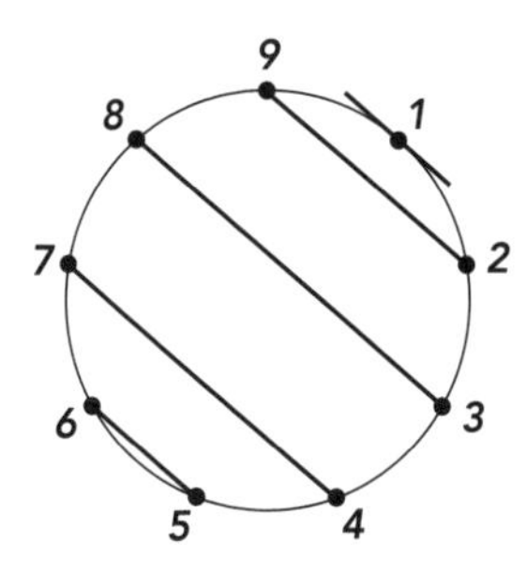

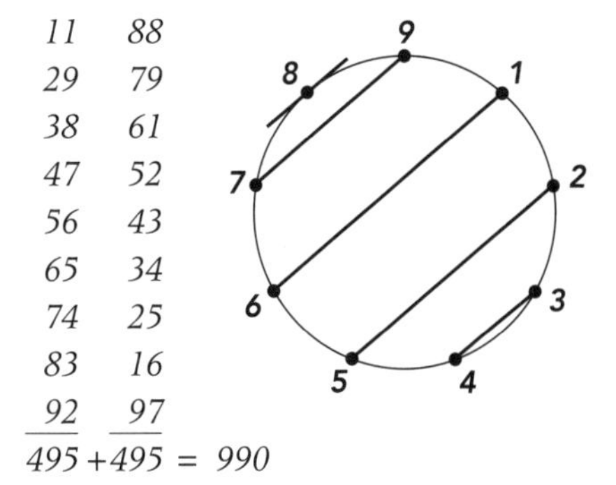

11	*88*
29	*79*
38	*61*
47	*52*
56	*43*
65	*34*
74	*25*
83	*16*
92	*97*

495 + 495 = 990

The Magic Sigma Square

3	1	5
8	6	4
7	2	9

Each row or column or diagonal adds up to Σ 9.

The square also reveals patterns that 'magically' add up to either, Σ 3, Σ 6, Σ 9. (Add the numbers represented by dots in the figures shown on the facing page).

But where does this square come from?

If we take the First Magic Square shown on page 52, add 1 to each number, and make 10 = 1 (for 1 + 0 = 1) and re-arrange, then we get the magic sigma square.

See what happens when the numbers in the square are raised to the second, third, fourth powers and so on and reduced to their sigma values. Compare the answers you get with the values running round the circumferences of the Power Mandala!

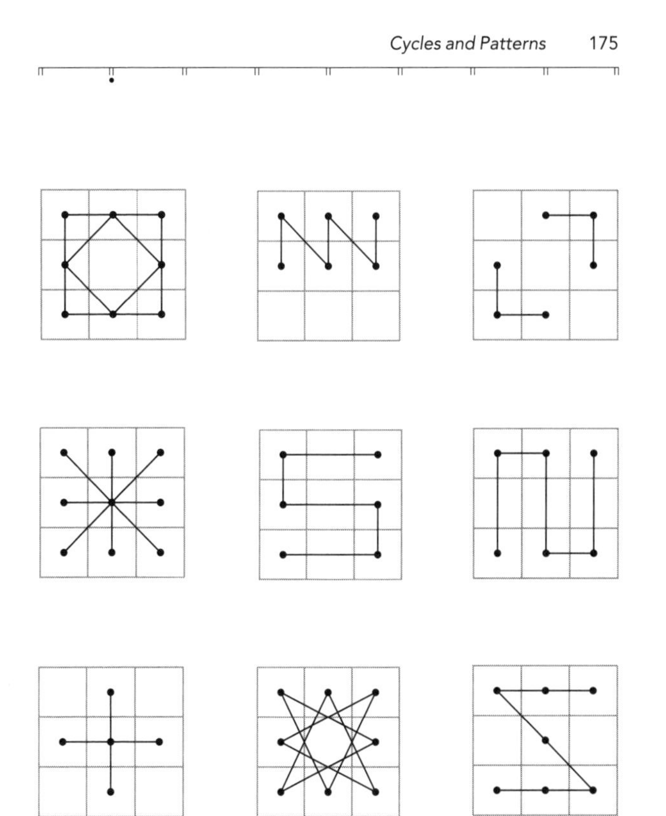

Calculator pad keys

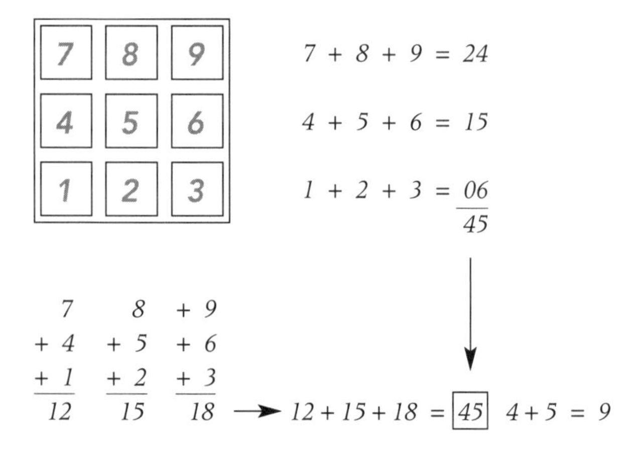

Relationship of the sigma circle to calculator pad keys

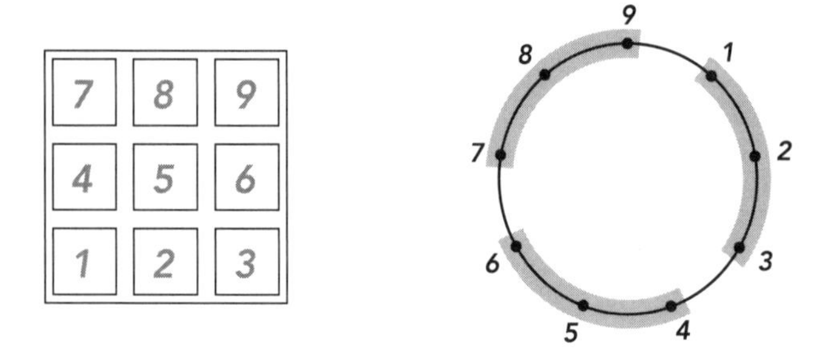

$$(1 \times 2 \times 3) + (4 \times 5 \times 6) + (7 \times 8 \times 9) = 630 \ (\Sigma 9)$$
$$(1 \times 4 \times 7) + (2 \times 5 \times 8) + (3 \times 6 \times 9) = 270 \ (\Sigma 9)$$
Adding up both rows $= 900 \ (\Sigma 9)$

Multiplying corners

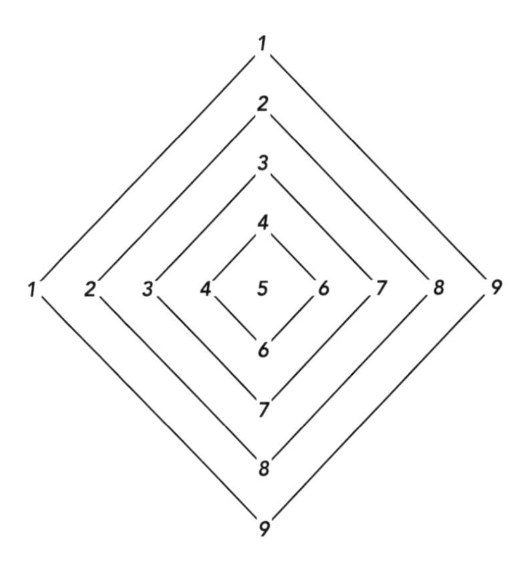

1 × 1 × 9 × 9 = 81
2 × 2 × 8 × 8 = 256
3 × 3 × 7 × 7 = 441
4 × 4 × 6 × 6 = 576

Add the centre + 005
1359

1 + 3 + 5 + 9 = 18 = Σ9

$$
\begin{array}{ccccccccc}
989^2 & - & 898^2 & = & \Sigma 1 & - & \Sigma 4 \\
878^2 & - & 787^2 & = & \Sigma 7 & - & \Sigma 7 \\
767^2 & - & 676^2 & = & \Sigma 4 & - & \Sigma 1 \\
656^2 & - & 565^2 & = & \Sigma 1 & - & \Sigma 4 \\
545^2 & - & 454^2 & = & \Sigma 7 & - & \Sigma 7 \\
434^2 & - & 343^2 & = & \Sigma 4 & - & \Sigma 1 \\
323^2 & - & 232^2 & = & \Sigma 1 & - & \Sigma 4 \\
212^2 & - & 121^2 & = & \Sigma 7 & - & \Sigma 7 \\
191^2 & - & 919^2 & = & \underline{\Sigma 4} & - & \underline{\Sigma 1} \\
 & & & & \Sigma 9 & - & \Sigma 9
\end{array}
$$

The above pattern is in the form of a continuous loop of the numbers 1 – 9 as repeated below.

Each leg is a rotation around the sigma circle, first anticlockwise, then clockwise and so on.

The net effect on the arithmetic is zero!

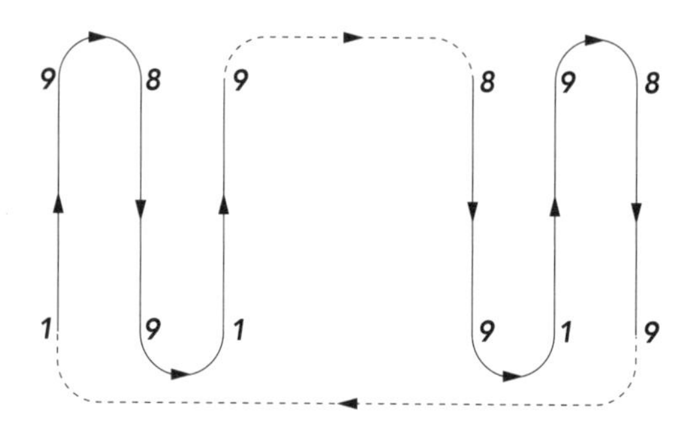

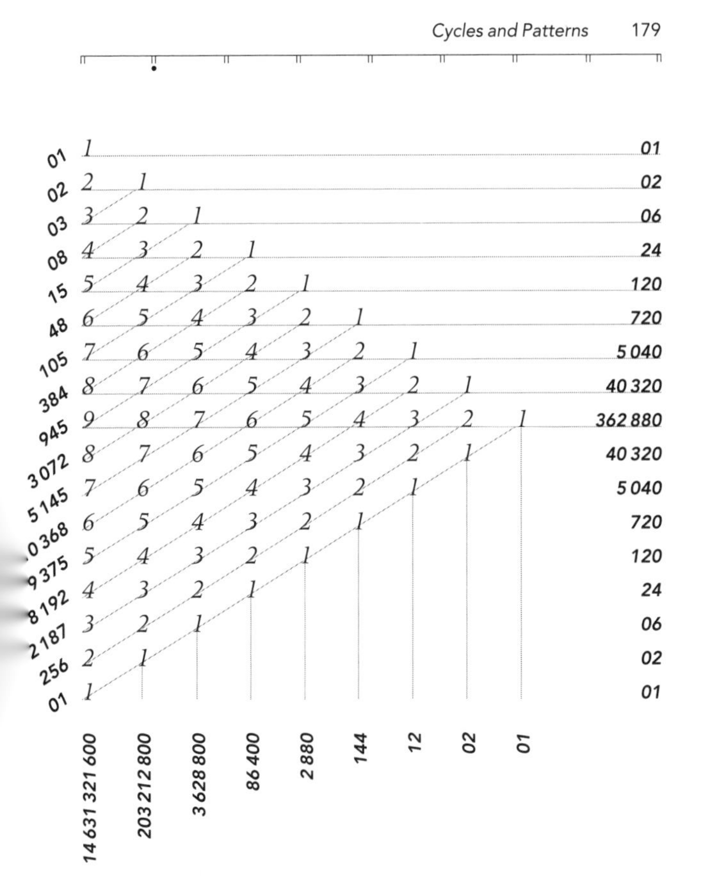

The numbers in the array above are multiplied out with each other in the horizontal direction, then separately both in the vertical direction and along the diagonal. When the products of multiplication are all added up, their sum equals diagonally Σ3 (40 107), vertically Σ6 (14 838 252 639) and horizontally Σ9 (455 346) – the stepping stones of 9!

The Complements of Nine

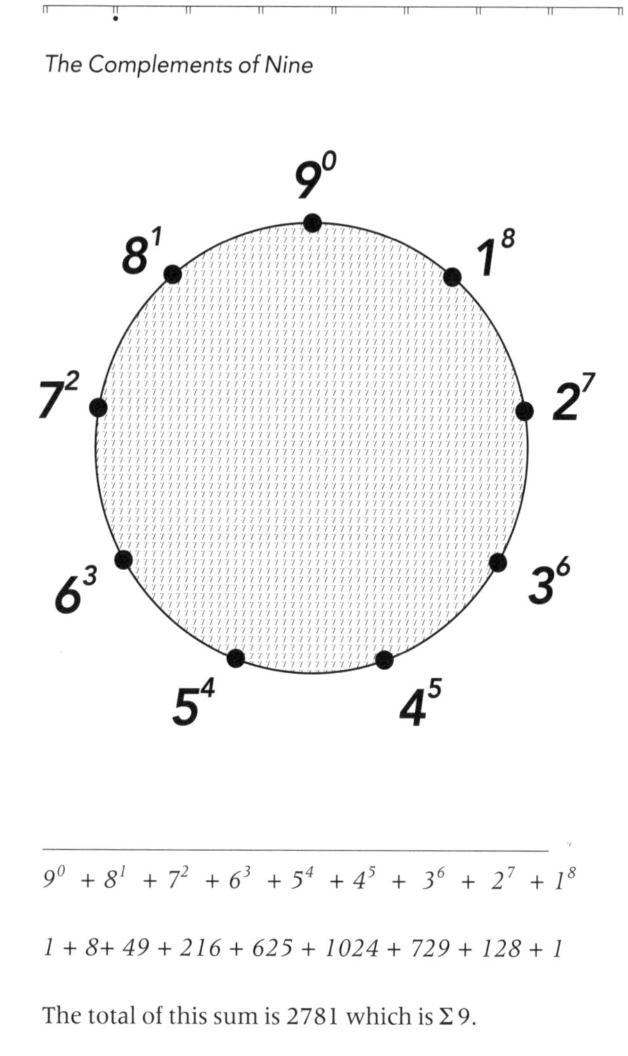

$9^0 + 8^1 + 7^2 + 6^3 + 5^4 + 4^5 + 3^6 + 2^7 + 1^8$

$1 + 8 + 49 + 216 + 625 + 1024 + 729 + 128 + 1$

The total of this sum is 2781 which is $\Sigma 9$.

Geometry and Symbol

The Star of David is made up of 6 triangles of 3 sides each, giving an 18-sided figure, i.e. $\Sigma 9$ straight line segments.

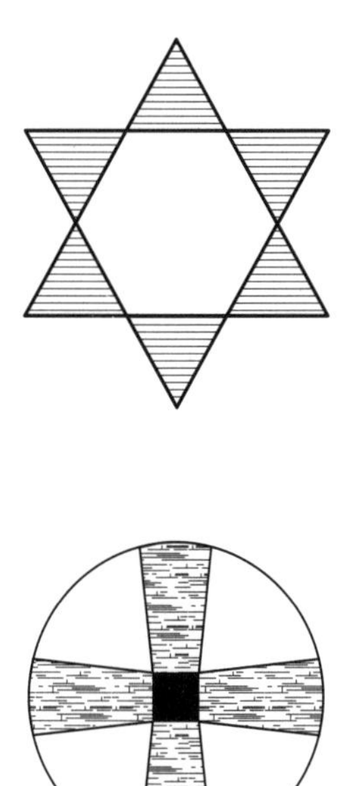

Nine regions make up the power of the Celtic Cross, that most ancient of Christian marks.

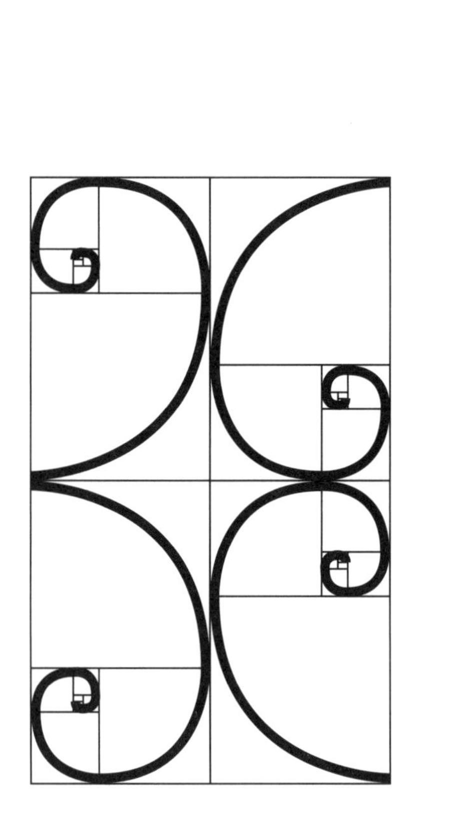

The Golden Section

The Golden Section is special in classical art and architecture. It is the particular ratio given by subdividing a line so that the greater part is to the lesser part as the whole is to the greater part.

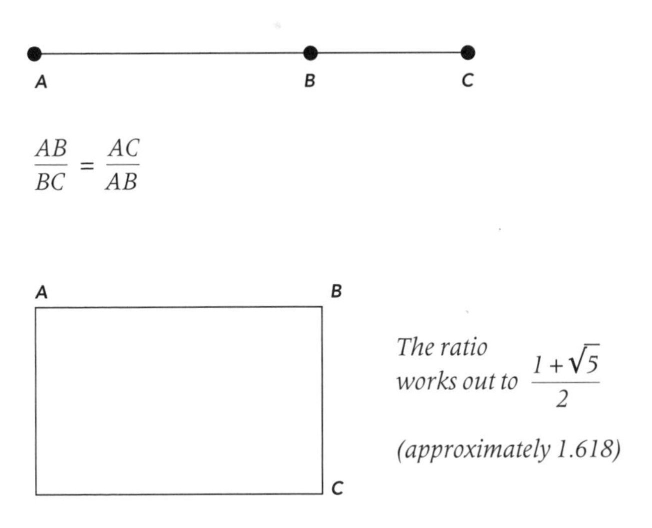

This was said to be the proportion found most pleasing in the rectangle and was used in the planning of the Parthenon and numerous other architectural works.

The golden ratio is also found in a triangle when it is isosceles (in which the two opposite sides are equal), but with a particular base angle.

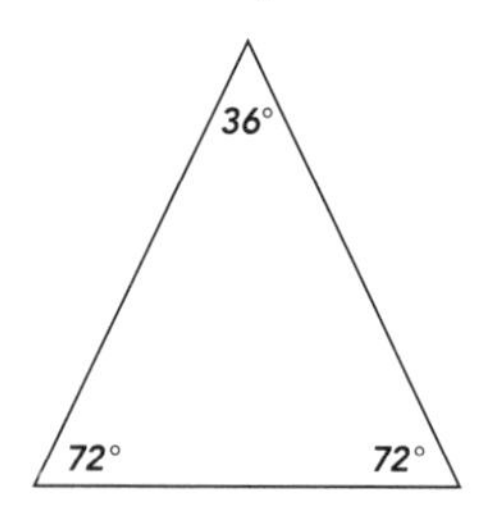

The ratio of the long side of the triangle to the shorter base is in the golden ratio when the base angle is 72°. This means the angle at the top vertex is 36°. Each angle of the Golden Triangle sums to nine!

The Greeks believed that at the heart of perfect proportion was the golden ratio and is it not pleasing to find number nine buried in the works, to those of us who chase its tail?

The unique property of the golden rectangle is that if a square is taken away from it, the remaining part is a rectangle, which has the same proportions to the original rectangle. That means it is golden as well. The process continues, with each removal of a square, leaving another golden rectangle. Smaller and smaller we go and if the points of the construction are joined together in a smooth curve, we enter a spiral, winding down into its vanishing point, the hidden eye.

The eye of the spiral is of course the shape of a vanishing 9.

Geometry is cornered around 9.

The triangle is a line that bends twice to close on itself, enclosing en route a total of 180° in its internal angles.

A square is a line that is more formal in its bending around corners, enclosing each time a right angle so that the sum of the internal angles is $4 \times 90° = 360°$.

A circle is a line that bends round continuously, about a centre point, enclosing the angle of 360°.

In general, any regular polygon encloses a total internal angle of (2n-4) right angles where n is the number of sides of the polygon.

For a pentagon n = 5 Sum of internal angles = 540°
For a hexagon n = 6 Sum of internal angles = 720°
For an octagon n = 8 Sum of internal angles = 1080°
For a decagon n = 10 Sum of internal angles = 1440°
etc.

In all cases the digits in the sum of internal angles add up to nine or multiples of nine. In other words for regular polygons

---

$$\Sigma \text{ (sum of internal angles)} = \Sigma 9$$

The Spiral in the Golden Section

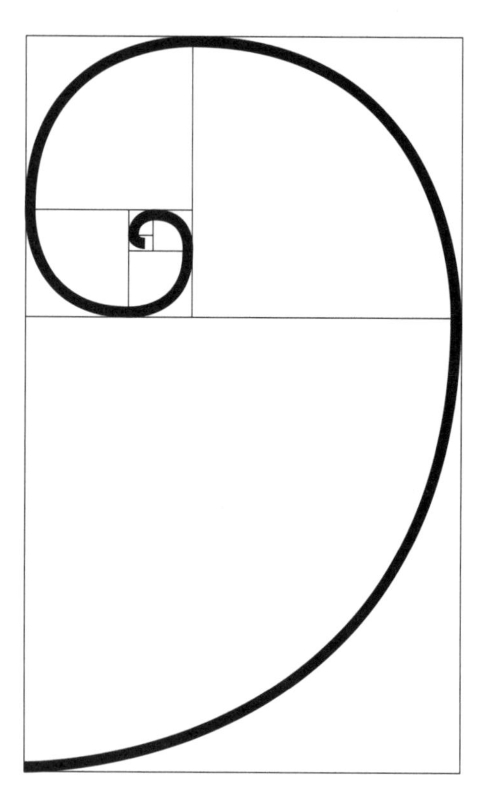

Pentagon and the Golden Section

Any diagonal of a regular pentagon in ratio to its side is golden:

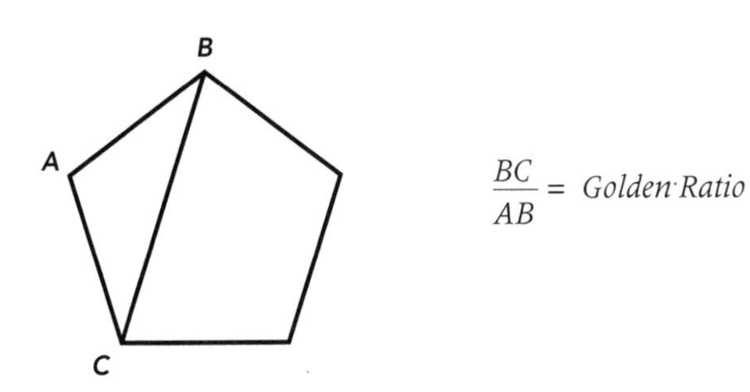

Extending any side of a pentagon gives the golden ratio:

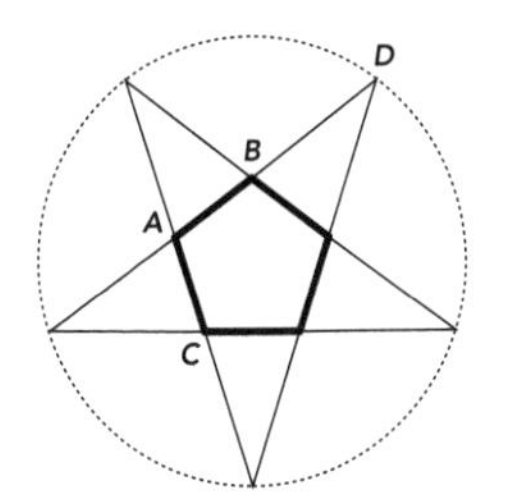

No wonder the pentacle is deemed to be magic!

Triangles

Pascal's Triangle is a famous mathematical contrivance that is fascinating and surprising. It is an array of numbers, built up by adding two adjacent values to get the next number below. For example 3 + 3 gives 6, or 6 + 15 gives 21. The array starts at the top with 1 and is bordered on each side by ones.

A pyramid is built up of numbers with very special properties.

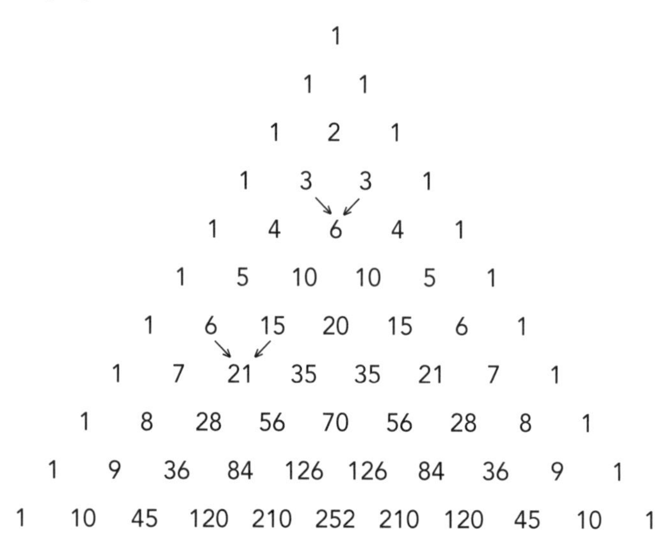

Amongst many other things Pascal's Triangle is a probability chart.

If two coins are tossed there is only one chance of getting both heads or tails but there are two ways of getting one head and one tail. The coefficients 1, 2, 1 are exactly the values of row 2 in Pascal's Triangle, if at the apex of the triangle, we take the single value 1 as row 0.

Heads Heads Heads Tails Tails Heads Tails Tails

Row 3 of Pascal's Triangle gives the probabilities when tossing three coins, and so on. But getting back to row 2 of the Triangle, the values 1, 2, 1 are also the coefficients of the expansion of the algebraic series.

$$(1+x)^2 = 1 + 2x + 1x^2$$

The expansion $(1 + x)^3$ will have coefficients from row 3 of the triangle, which are 1, 3, 3, 1.

$$(1+x)^3 = 1 + 3x + 3x^2 + 1x^3$$

There are many other interesting properties to this fantastic piece of organisation in numbers – and the keen reader must look to more mathematical texts on Pascal's Triangle for further elaborations.

When the sigma code is applied to Pascal's Triangle a curious thing happen. Unlike the numbers in the triangle which keep growing exponentially, I find that beneath this pattern of growth there is a hidden consistency.

Each row, when added up horizontally, reduces to a single value as part of a set sequence of six values:

1 – 2 – 4 – 8 – 7 – 5

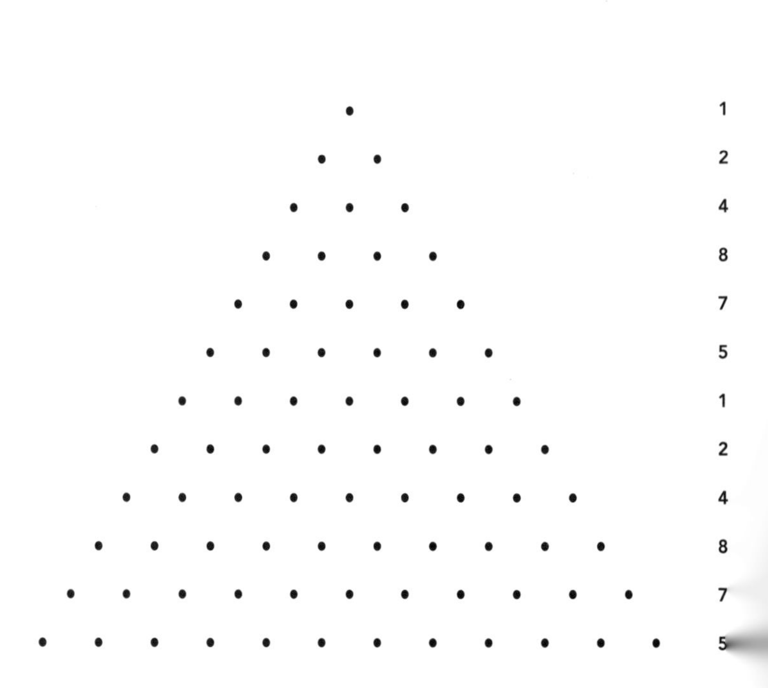

This is a real surprise; a hidden conspiracy going on in the hugely versatile and inventive triangle under the surface. What is rather special is that the values of the sequence 1 – 2 – 4 – 8 – 7 – 5 add up to 27 which gives 2 + 7 = 9.

I looked into sixty rows of the triangle, where the numbers become really gigantic, but each row still reduced to the same sigma sequence above, repeating after every six rows. Is this a coincidence or just extra-ordinary?

But this sequence is also found in radiating arm [2] of the power circles in Part I of this book. In other words, the sequence 1 2 4 8 7 5 is the expansion of the power of 2.

$2^0 \quad 2^1 \quad 2^2 \quad 2^3 \quad 2^4 \quad 2^5$

The sigma code shows something in this triangle that is not obvious at first glance.

I found further resonance in the triangle with the stepping stones of 9, viz those of 3 and 6.

Each band of *six* rows of the triangle contains a number of entries (here we are not considering the actual values, just their location) that sum up to *three* when reduced to the sigma code. For example the first six rows have 21 entries (Σ 21 = 3) and the next six rows have 57 entries (Σ 57 = 3), and so on.

But the loveliest touch of all is that the repeating sequence of 1 – 2 – 4 – 8 – 7 – 5 when plotted on the sigma circle yields a symmetric plot about the vertical axis.

This sequence that repeats as the triangle grows, shows that under the surface something is conserved; a secret energy under the obvious multiplicity and growth of the actual values of the triangle. And conservation has its reward – on the sigma circle the sequence has symmetry.

The intrigue is that the sequence 1 2 4 8 7 5 has the same numbers as the cyclic number 1 4 2 8 5 7, just the order is different. See "Cycles".

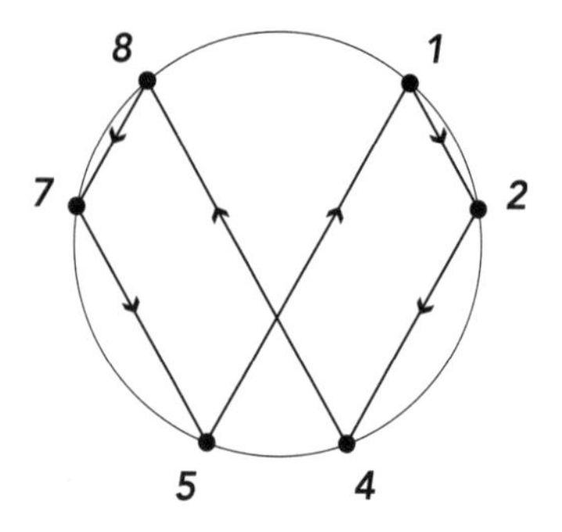

The listings of the sums of the first 36 rows of Pascal's Triangle with the relevant sigma values.

Rows 1–18	Σ Values	Rows 19–36	Σ Values
1	1	262144	1
2	2	524288	2
4	4	1048576	4
8	8	2097152	8
16	7	4194304	7
32	5	8388608	5
64	1	16777216	1
128	2	33554432	2
256	4	67108864	4
512	8	134217728	8
1024	7	268435456	7
2048	5	536870912	5
4096	1	1073741824	1
8192	2	2147483648	2
16384	4	4294967296	4
32768	8	8589934592	8
65536	7	17179869184	7
131072	5	34359738368	5

The sequence 1 – 2 – 4 – 8 – 7 – 5 which sums to Σ 9 repeats throughout the triangle.

Older Version of Pascal's Triangle

The Chinese used such a triangle hundreds of years before the French mathematician – though they organised the triangle differently – as a right angled triangle.

1 1

1 2 1

1 3 → 3 1

↓

1 4 6 4 1

1 5 10 10 5 1

1 6 15 20 15 6 1

1 7 21 35 35 21 7 1

1 8 28 56 70 → 56 28 8 1

↓

1 9 36 84 126 126 84 36 9 1

1 10 45 120 210 252 210 120 45 10 1

Each number is the result of a 'dog leg' addition:

$3 + 3 = 6$ *or* $70 + 56 = 126$ *etc.*

The Diamonds in Pascal's Triangle

Beneath the rapid growth of numbers in Pascal's Triangle we tracked a secret code at work – each six rows giving a repeating sequence of numbers whose sigma values added up to nine.

But what happens if we write out Pascal's Triangle in sigma values (so that 6 + 4 = 10 is written as 1, and 20 + 15 = 35 is written as 8 and so on)?

In fact we do not need to do even these conversions, the triangle can be constructed from the outset by just using the logic of reducing any new entry to the relevant sigma value.

Now if a pattern of diamonds is set up in the triangle, their sigma values add up to 9 when going across the triangle, except for the first entry.

See the triangles overleaf.

Pascal's Triangle with Sigma Values

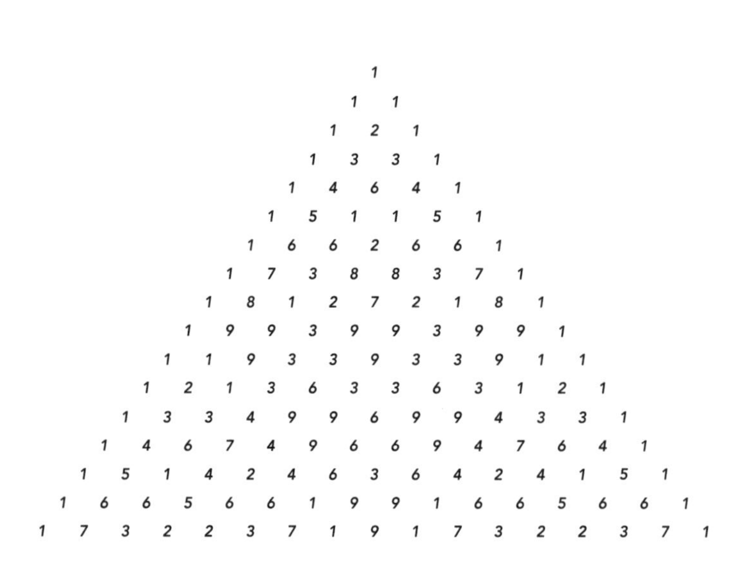

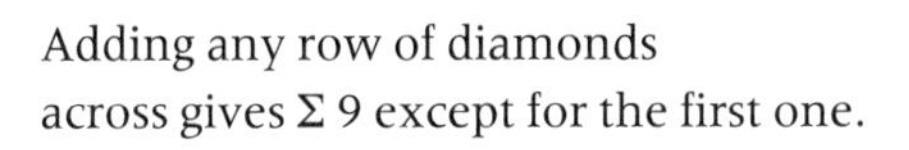

Adding any row of diamonds
across gives Σ 9 except for the first one.

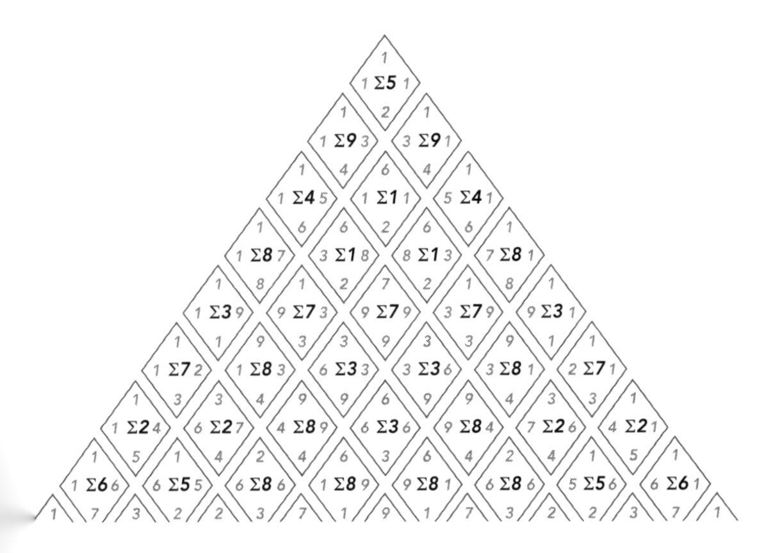

Primes

In the world of numbers some numbers have very special properties. They are the prime numbers. They are unique in that they cannot be factored. Numbers like seven, eleven, seventy-nine, one hundred and one, are such numbers. All other numbers are called composite and can be factored down into primes. For example 276 and 390 have the following factors, which are prime:

$$276 = 2 \times 2 \times 3 \times 23$$
$$390 = 2 \times 3 \times 5 \times 13$$

Ultimately, primes are the 'atoms' of numbers.

What is much celebrated about the primes is their distribution. It appears to be random, starting with 1 and continuing:

1 2 3 5 7 11 13 17 19 23 29 31 37 . . .

Mathematicians have tried over the centuries to predict this unfolding sequence with a formula, but have failed. The only successful solution to date has been a default method, invented by Eratosthenes, of writing out all the integers in sequence and crossing off successively the multiples of 2, 3, 4, 5, 6, etc. The numbers left after this 'sieving' process are, miraculously, the primes! A sample of the method from numbers 2 to 52 is shown below.

2 3 ~~4~~ 5 ~~6~~ 7 ~~8~~ ~~9~~ ~~10~~ 11 ~~12~~ 13
~~14~~ ~~15~~ ~~16~~ 17 ~~18~~ 19 ~~20~~ ~~21~~ ~~22~~ 23 ~~24~~ ~~25~~ ~~26~~
~~27~~ ~~28~~ 29 ~~30~~ 31 ~~32~~ ~~33~~ ~~34~~ ~~35~~ ~~36~~ 37 ~~38~~ ~~39~~
~~40~~ 41 ~~42~~ 43 ~~44~~ ~~45~~ ~~46~~ 47 ~~48~~ ~~49~~ ~~50~~ ~~51~~ ~~52~~

The study of primes is a fascinating subject because one is always trying to find if there is a hidden pattern or some trace of conformity.

As the sigma code looks under the surface of arithmetic I thought it would be an ideal tool for finding such traces, if they existed. And what I found under the x-ray eye of the code is a special relationship that the primes seem to have to the numbers 3 – 6 – 9.

Beyond number 3, all primes add up to one of the following sigma code values

1 2 4 5 7 or 8

but their digits never seem to add up to 3, 6 or 9. It would appear that the character of 9 is avoided.

Further analysis however shows that this is not the case and that actually number nine is buried deep in the works, a secret, behind a secret.

To find out more, I raised the prime numbers to successive powers and tabulated their sigma code values. And the answers still avoided being 3, 6 or 9; but when I went further and added up these products, mysteriously the character of nine began to appear: as 9 itself or in the guise of one of its stepping stones – 6.

For example the prime number 13 raised from the 1st power to the 6th power gives successive sigma values as:

Prime Power	$(13)^1$	$(13)^2$	$(13)^3$	$(13)^4$	$(13)^5$	$(13)^6$
Value	*13*	*169*	*2197*	*28561*	*371293*	*4826809*
Sigma Value	*4*	*7*	*1*	*4*	*7*	*1*

There is no 3, 6 or 9 here. But if these values are added up horizontally, 4 + 7 + 1 + 4 + 7 + 1 they add up to 24 and 2 + 4 = 6.

Let us carry the investigation further; prime numbers 17, 19, 23, give the following sigma power sequences:

	$(\ )^1$	$(\ )^2$	$(\ )^3$	$(\ )^4$	$(\ )^5$	$(\ )^6$		
17	*8*	*1*	*8*	*1*	*8*	*1*	→	*9*
19	*1*	*1*	*1*	*1*	*1*	*1*	→	*6*
23	*5*	*7*	*8*	*4*	*2*	*1*	→	*9*

Adding the results horizontally leads to the values shown in the extreme right hand column, an unfolding sequence of 6 and 9.

As this working continues, the columns of numbers in the sixth column show a remarkable consistency. All primes to the sixth power, except for number 3, have digits that sum up to unity! This is a shock to the system. How do these random numbers suddenly show such conformity? And it does not stop there; this fact holds for the 12th power, the 18th power and so on.

Another way of saying this is that at every sixth self-multiplication prime numbers come home as it were, in a great levelling off, with their digits adding up to unity.

On the following page is a listing of the prime numbers from 1–109 giving the sigma values of the primes themselves and their successive powers.

The far right-hand column, which is the sum of the first six powers, shows the hidden organisers 6 and 9 at work.

Primes have many secrets
but what the sigma code reveals must rank
as one of the strangest.

$$\Sigma\ (N\ prime)^6 = 1$$

Chart of Sigma Values of Primes

Prime numbers raised to successive powers, with their associated sigma value, tabulated up to the sixth power.

Prime Number N	$(\Sigma N)^1$	$(\Sigma N)^2$	$(\Sigma N)^3$	$(\Sigma N)^4$	$(\Sigma N)^5$	$(\Sigma N)^6$		Adding columns one to six
1	1 +	1 +	1 +	1 +	1 +	1	➜	6
2	2	4	8	7	5	1		9
3	3	9	9	9	9	9		3
5	5	7	8	4	2	1		9
7	7	4	1	7	4	1		6
11	2	4	8	7	5	1		9
13	4	7	1	4	7	1		6
17	8	1	8	1	8	1		9
19	1	1	1	1	1	1		6
23	5	7	8	4	2	1		9
29	2	4	8	7	5	1		9
31	4	7	1	4	7	1		6
37	1	1	1	1	1	1		6
41	5	7	8	4	2	1		9
43	7	4	1	7	4	1		6
47	2	4	8	7	5	1		9
53	8	1	8	1	8	1		9
61	7	4	1	7	4	1		6
67	4	7	1	4	7	1		6
71	8	1	8	1	8	1		9
73	1	1	1	1	1	1		6
79	7	4	1	7	4	1		6
83	2	4	8	7	5	1		9
91	1	1	1	1	1	1		6
97	7	4	1	7	4	1		6
101	2	4	8	7	5	1		9
103	4	7	1	4	7	1		6
109	1	1	1	1	1	1		6

The Hidden Organisers

Throughout the Movements of Nine the investigation has shown that despite its organising power, number nine seems to act as an innermost secret, not revealing itself at first.

Upon first analysis the sigma code values we come across are:

1, 2, 4, 5, 7 or 8

In Pascal's Triangle every six rows has the sequence 1 – 2 – 4 – 8 – 7 – 5. And in Cycles, the number that rotates itself under multiplication is 142857, a slight alteration to that of the Pascal's Triangle sequence. Yet again in looking at prime numbers, their sigma code value is either 1, 2, 4, 5, 7 or 8 (the number 3 is the one exception to this).

What is missing from these numbers in general are the values 3, 6 or 9. Yet as we have seen in the text, the character of nine ultimately does appear. It is as if the numbers three, six and nine remain hidden to serve as some kind of conduit or template through which the other numbers pass.

If the numbers 1, 2, 4, 5, 7 or 8 carry the motion of the sigma code and first announce a secret, then the stepping stones ③ – ⑥ – ⑨ must remain the innermost secret, to allow this to happen.

We can gain an insight into this behaviour when we look at the magic sigma square or the sigma circle.

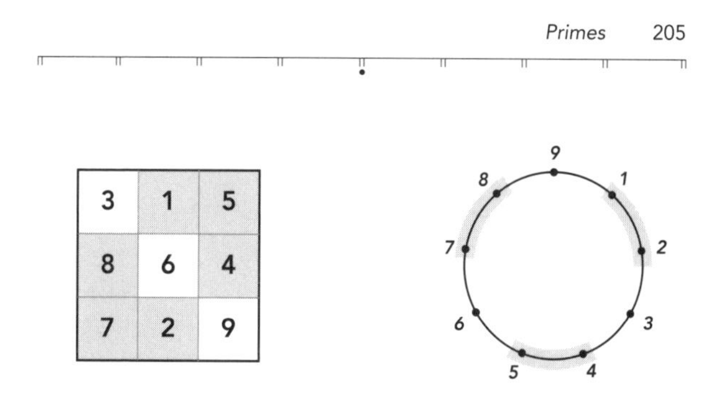

In the magic sigma square the numbers three, six and nine are on the diagonal like an organising principle, pushing the other numbers to each side of them. Similarly, in the sigma circle the numbers three, six and nine form a robust triangle with the other numbers to each side of this central anchor.

The numbers ③ – ⑥ – ⑨ seem to be both conduit and control in a secret world of numbers.

It is the principle of the active and the passive, the Yang and Yin, the active core that illuminates buried secrets.

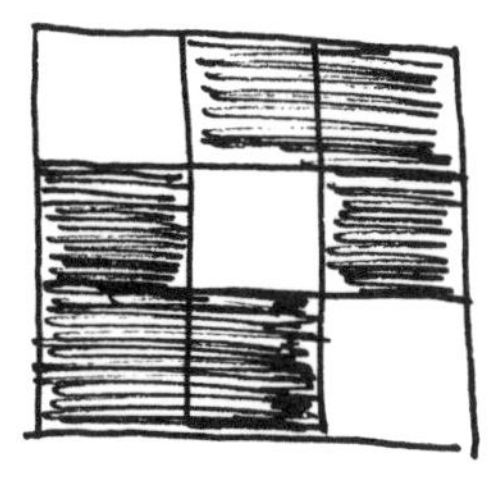

Equations

As revealed by the sigma code, the way number nine affects arithmetic can be written out in a set of strange equations, where N is the sum of the digits of any number:

$$\Sigma N + \Sigma 9 = \Sigma N$$
$$\Sigma N - \Sigma 9 = \Sigma N$$
$$\Sigma N \times \Sigma 9 = \Sigma 9$$
$$\Sigma N \div \Sigma 9 = \Sigma N \ (\textit{remainder})$$

In the case of prime numbers there is a very curious equation I found that results if a prime number is raised to the sixth power and its digits added up.

Then:

$$\Sigma (N\,prime)^6 = \Sigma 1$$

For details of how this happens see "Primes".

The above equations, strange as they seem, can be explained mathematically by using the following facts:
– number nine is equal to ten minus one

$$9 = 10 - 1$$

– and any number can be expressed as an algebraic equation of the form

$$ax^n + bx^{n-1} + \dots + (\,)x^1 + (\,)x^0$$

where a, b, etc. are integers 1, 2, 3, etc. and x is the relevant base of numbers.

Now in our decimal system x = 10, any two digit number, say, can be expressed as:

$$ax^1 + bx^0$$

or more simply as $ax + b$ *because* $x^1 = x$ *and* $x^0 = 1$

The sigma value of this number is the same as its coefficients so that:

$$\Sigma\, number = (a + b)$$

Now if 9 (which is 10 – 1 or in algebraic notation x – 1) is added to the original number, the new number is:

$$(ax + b) + (x - 1)$$

Adding like coefficients the new number is:

$$(a + 1)x + (b - 1)$$

The sigma value of this number is the sum of its coefficients.

$$\Sigma \textit{ New Number} = a + 1 + b - 1 = a + b$$

That is the same sum as the original number before the addition of nine. In other words adding nine makes no difference to the sum of the digits. This proof for addition is the simplest. For subtraction, the logic is similar but you need to watch carry-overs. Multiplication is also similar, and again watch the carry-overs. But for division, the reader needs to know about congruences.

A number like 10 divided by 9 is said to be congruent to 1 and written:

$$10 \equiv 1$$

and is stated as "10 is congruent to 1 modulo 9". This is a more advanced branch of number theory and the reader who wants to understand these proofs in detail should consult one of the books listed in the bibliography.

No doubt for the pure mathematician the algebraic explanations are succinct, simple and elegant; but what is fun is to find a graphical way of explaining these abstractions.

To most people mathematics is a taboo area, an area they do not feel comfortable with. That is a pity, because mathematics is beautiful; the delight being in chasing purely mental constructs.

The strange power of number nine may be written out in only a few lines of algebra yet there is something more, something much more interesting when a picture emerges from out of the abstraction.

The sigma circle is such an invention. In one orbit it explains the unique property of nine. And in some ways it is more powerful than the mathematics, because the picture penetrates deeper into the area behind our rational mind. The saying goes that a picture says much more than words; in that sense the sigma circle and the greater inspiration of its Mandalas speaks volumes, sharpening our intuition beyond a number itself into patterns and characters. Whereas mathematics describes, a picture inspires. Equations demand our intellect, but the Mandalas demand curiosity and our poetry.

From pure mathematics we can see that nine owes its character to the unique position it has in our decimal system. But number nine has also travelled far beyond these computational confines. From another route, through the mists of time and long before the invention of our decimal system, number nine grew and developed its own archetypes and mystery.

Vishnu took three strides to create the universe. Three times three is potency magnified beyond measure. That beginning of the great rebirth of the seasons, the winter solstice, is in the ninth house of the Zodiac; and the infant takes nine months to be born. Unaided by any mathematical proofs number nine grew its own equations in the human imagination. And Allah was honoured and blessed and was called by ninety nine names.

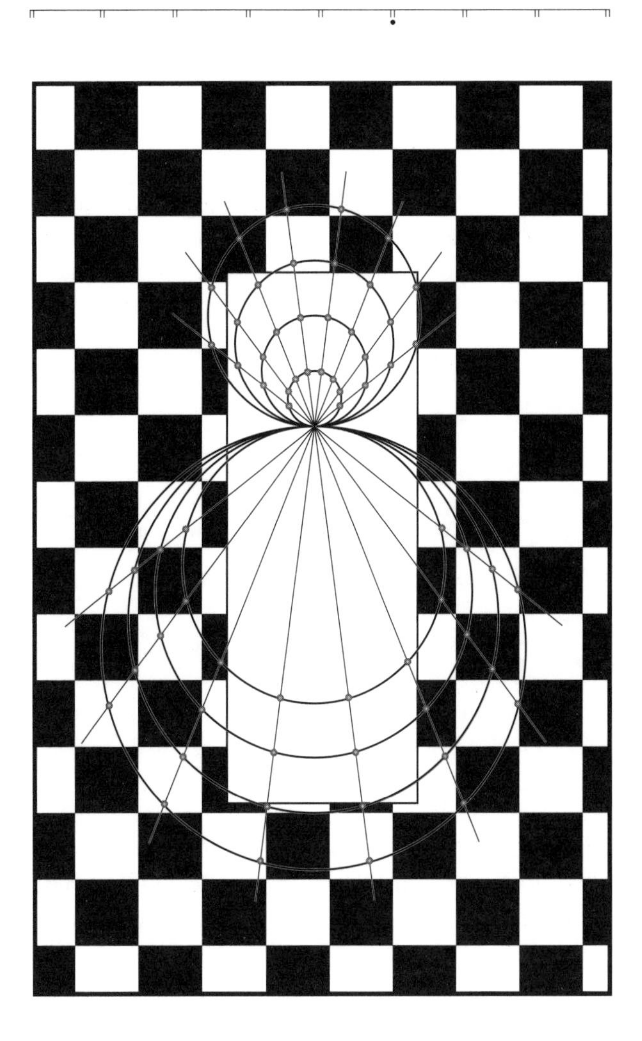

Proverbs and Charms

A stitch in time saves nine

Possession is nine tenths (or nine points) of the law

Dressed up to the nines

At the ninth hour

Right as ninepence

Nine times out of ten

Nine sheets to the wind

Ninepence to the shilling

Nimble as ninepence

To look nine ways

Nine days wonder

Nine tailors make a man

To be on cloud nine

When constructing a magic circle it should be nine feet in diameter.

Fairies come out to dance when you put nine grains of wheat on a four leaf clover.

To bring good luck the ABRACADABRA is worn as a charm around the neck, for nine days, and then thrown away.

To charm away a sprained ankle you put nine knots on it, made out of black wool.

To charm away an itch, a left handed man should whirl a black cat round his head three times, make an ointment from nine roasted barleycorns and nine drops of blood from the cat's tail and apply this while walking round the person with the itch three times, calling on the Trinity.

To bring grief to an enemy measure a corpse with a piece of rope three times, from three different points ($3 \times 3 = 9$): from the elbow to the longest finger, from the shoulder to the longest finger, and from head to toe. Then 'capture' the person you want to inflict the grief upon, and measure him or her in the same way as you did the corpse!

Naming

The 99 Names of Allah

Allah	*The Mighty*	*The Independent*
The Compassionate	*The Forgiving*	*The Powerful*
The Merciful	*The Grateful*	*The Dominant*
The King/Sovereign	*The High*	*The Giver*
The Holy	*The Great*	*The Retarder*
The Source of Peace	*The Preserver*	*The First*
The Giver of Faith	*The Protector*	*The Last*
The Overall Protector	*The Reckoner*	*The Manifest*
The Strong	*The Sublime*	*The Hidden*
The Almighty	*The Bountiful*	*The Governor*
The Majestic	*The Watcher*	*The High Exalted*
The Creator	*The Responsive*	*The Righteous*
The Maker	*The Infinite*	*The Relenting*
The Fashioner	*The Wise*	*The Forgiver*
The Great Forgiver	*The Loving*	*The Avenger*

The Dominant	*The Glorious*	*The Compassionate*
The Bestower	*The Resurrector*	*The Ruler of the Kingdom*
The Provider	*The Witness*	*The Lord of Majesty and Bounty*
The Opener	*The True*	*The Equitable*
The All-Knowing	*The Advocate*	*The Gatherer*
The Restrainer	*The Most Strong*	*The Self-Sufficient*
The Extender	*The Firm*	*The Enricher*
The Humbler	*The Patron*	*The Bestower*
The Exalter	*The Praiseworthy*	*The Withholder*
The Empowerer	*The Numberer*	*The Propitious*
The Humiliator	*The Commencer*	*The Distresser*
The All-Hearing	*The Restorer*	*The Light*
The All-Seeing	*The Giver of Life*	*The Guide*
The Judge	*The One Who Gives Death*	*The Eternal*
The Just	*The Living One*	*The Everlasting*
The Kindly One	*The Self-Subsisting*	*The Heir*
The Gracious	*The Perceiver*	*The Guide to the Right Path*
The Clement	*The One*	*The Patient*

Symbol

3 x 3

In the field of all possibilities the Wise Ones decided to place a system of markers. In the blankness there would be direction and reference, how else was the race to go forward?

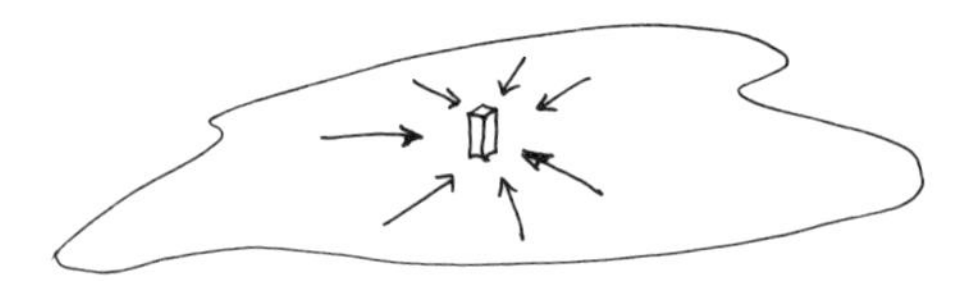

They put down one marker in the open landscape. Everyone ran up to the great stone column shrieking at its newness. They swarmed around it but could not judge what else to do. In this infinite plane the stone column, a point, became an infinity itself: it was everything. The point soon became pointless. So the Wise Ones tried two points.

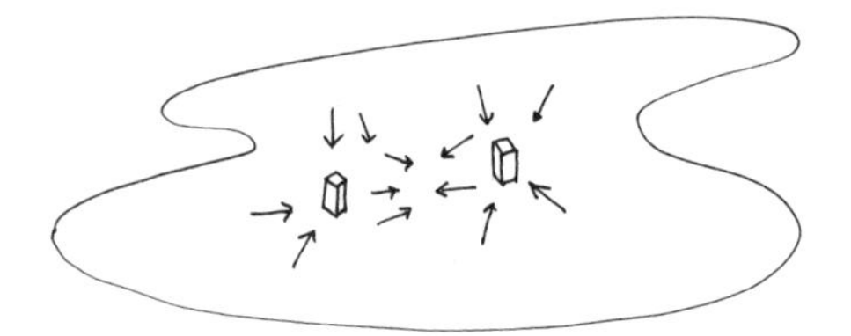

Some of the people went to the old stone column, the others ran to the new shining column, glinting in the land. Some Clever Ones connected one point to the other in a straight line and rushed up and down, between the points. But that was all they could do and soon bumped into each other. Some claimed one point as theirs. Why they chose one stone column and not the other they could not say, but each group thought their marker significant and the most important one. The groups began to shout at one another, claiming their marker was the right one. Fights broke out. So the Wise Ones decided to try one more time and threw down another marker, in a great tearing sound, through the air.

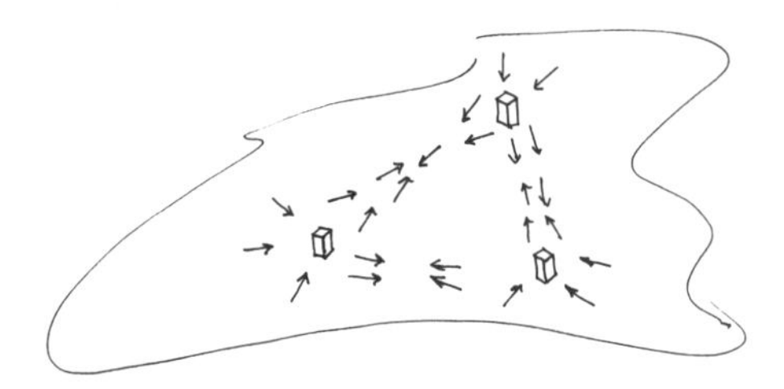

Now there were three markers. Suddenly a kind of democracy set in. People could not stand opposed one to the other. It was more complex. There had to be intrigues. Three points suddenly gave the people a sense of space, they could set up enclosures and boundaries. They built fences. The people got used to each other and soon the race began to think as one.

And more people came travelling through the blank spaces. The crowd grew into a multitude. In relation to the blank open space the three markers became an enclosure, a closed point, the people swarming like ants around it, hanging outside the imaginary walls, knowing not what else to do.

But the Great Ones wanted a distribution of evenness in the open spaces and they threw down another three markers. There was a rush from all those who camped outside the walls of the old triangle into the new region. But the two regions, of three points each, opposed one other; it was as if they glared across the divide at each other. What happened was that those in each region became an entity, the space they inhabited shrunk in their mind, as if to a point. And so point opposed point again. War broke out. The Elders moved swiftly then, streaming three more markers into the wide blank plane; and this gave them what they wanted, an evenness of distribution, a wider peace. No one camp could fight another – the cross lines grew too complicated. They argued and called each other names but no two could agree to fight one another, so peace prevailed, crossed in three times three ways.

The Elders called the idea wonderful; and three pillars, in groups of threes, all nine of them, marked the land to bring it out of the wilderness and blankness.

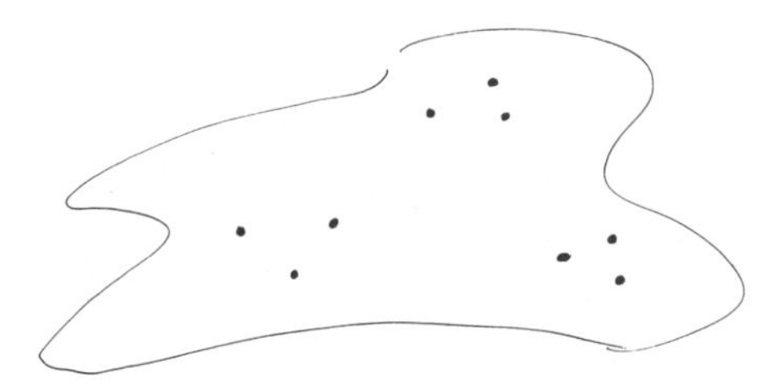

Enjil's Mark

The numbers 1–9 are symbols and exert a powerful hold on the imagination – they act as some kind of bridge to other worlds. Because they are inviolate and basic, they become the hidden steps to another kind of understanding, to prophecy and divination.

In researching this book I came across the Cabalah, and the faith it had in Numerology, which used the numbers 1–9 to divine a person's character. I tried the method on Enjil and got the number 5, for the Self or Life Number which is the number that describes the essence of a character. It is also the number for magic in Numerology. What was curious was that my first name also came to this number.

I then tried out another aspect, the Personality number, which is given by checking the consonants in a name. The letters N J L in 'Enjil' added up to nine. I was quite stunned and delighted. But so did the consonants in my first name: C C L.

Was I destined to write this book? Had Enjil called me up?

Was there another level operating here, a deeper intuition, that guided me towards Enjil?

Sometimes I think with a slight touch of fear that perhaps he was out there waiting to come in – an idea looking my way, hovering. Then in the yellow light of the moon, in one night of asking strange questions, a ray enters my mind and illuminates the path. Enjil invades. I feel as if he has invented a new pair of eyes in me so that I can see clearer, undoing the clutter and the learning I carry around like heavy baggage and I learn to count again in simple ways, 1, 2, 3, ...

The Last Movement

When I began looking into numbers I had no idea that number nine was such fertile ground. I read the books and followed certain well-worn explanations but could not find what I really wanted, a picture, a diagram. In mathematical books, number nine remained a mental construction. But in the primitive belief that numbers needed shape I tried to create a physicality around the idea. Then Enjil showed it to me, the sigma circle.

If I arranged the numbers one to nine, in a circle, suddenly everything fell into place. I had looked through all the books I had on numbers and searched the contents of others on bookshelves – but did not find a graphic explanation for the special behaviour of nine. And when I found it, there were the pictures for the other numbers as well, one to eight, the whole thing held tightly by the character of nine. It was simple and quite beautiful.

But Enjil had come to me previously.

I was working on another puzzle, on patterns, for another book. I was looking into how to connect up eighteen fixed points, laid out in 3 × 6 rows, in as many ways as possible. Just as I was exhausting the possibilities thrown up by a logical approach I thought of the games I had played in childhood, skipping along large stones by riverbanks, constructing special roadways beside the babbling water. And suddenly a spirit sprang to life, a boy-wonder of mathematics releasing a whole new set of patterns in my mind. Like a guardian angel, he skipped over the stones giving an answer to the puzzle I was looking into. That boy's name was, of course, Enjil. So when I was struggling again to find a picture and a pattern for number nine, I thought of Enjil and invoked that spirit of invention. That is when the answers came; the sigma circle of nine numbers, the other orbits of the Mandalas, and the meditations of moving patterns. I drew upon the woman Soma who lives in the moon to inspire me, for in Hindu myth she is the essence of wisdom, the drink of the Gods, the nectar of a divine inspiration. She is magic. And her Greek namesake, Sophia, is also an essence, one of heavenly wisdom; who better than these spirits to inspire Enjil and me into action?

After the Mandalas were found, looking for the movements of nine and searching for the hidden sequences buried deep in mathematical constructions, became absorbing research.

Prime numbers suddenly conformed under the spell of the sigma code, despite their unique and stand-alone indivisible nature. Nothing I had learnt about primes had led me to believe that there was anything to them but their oddness; yet the code opened a door to speculation about their behaviour.

Perhaps the surface eccentricities of the distribution of numbers, as with prime numbers, have inner and deeply buried movements of order, a gathering together of certain essences so that nothing is truly random. Perhaps for something to appear in our world and to be tracked by our minds, there must be some kind of hidden coherence, though that secret would be buried deep and generally unknown to us. The sigma code certainly points to this conjecture and makes us think of numbers in different and new ways. The sigma code and Enjil's story give both animation to numbers and new hope.

But the Mandalas also taught me something else: that "modern" need not mean "different", a fracture from the past. What is interesting in the Mandalas (pages 122 and 134) is that both diagrams, the classic centred one of symmetry and the strange skewed one, hold the same information; one static and concentric, the other swinging in and out, and dynamic.

From our pasts, the images of completeness we take in are symmetric; and perfection is said to radiate outwards from a point, and what surrounds a source is held at equal distances. That was how I constructed the first Mandala, spaced out, evenly, around the innermost sigma circle.

I was trapped by the static concept. I even called the diagram a Mandala, a meditation along Eastern lines of fixed centred thinking.

But to my mind, the Strange One, a modern Mandala for today, is even more exciting than the concentric one. The removal of symmetry lends a particular dynamic; the contemplation is still there but it is more informed, if anything, more focussed towards a greater concentration on the true balance point, the cross-over of all orbits.

What we learn from these diagrams is that a modern notation is not necessarily a break from the past. It is just a reworking of the same information in a wholly new way, more relational than absolute. And perhaps more accurate for that. We need not fear the apparent disjointed and twisted surfaces of the non-symmetric, it is just another way of looking, to my mind, a more accurate view perhaps of really what is going on around us. And that gives me hope, something to look forward to in the realm of ideas.

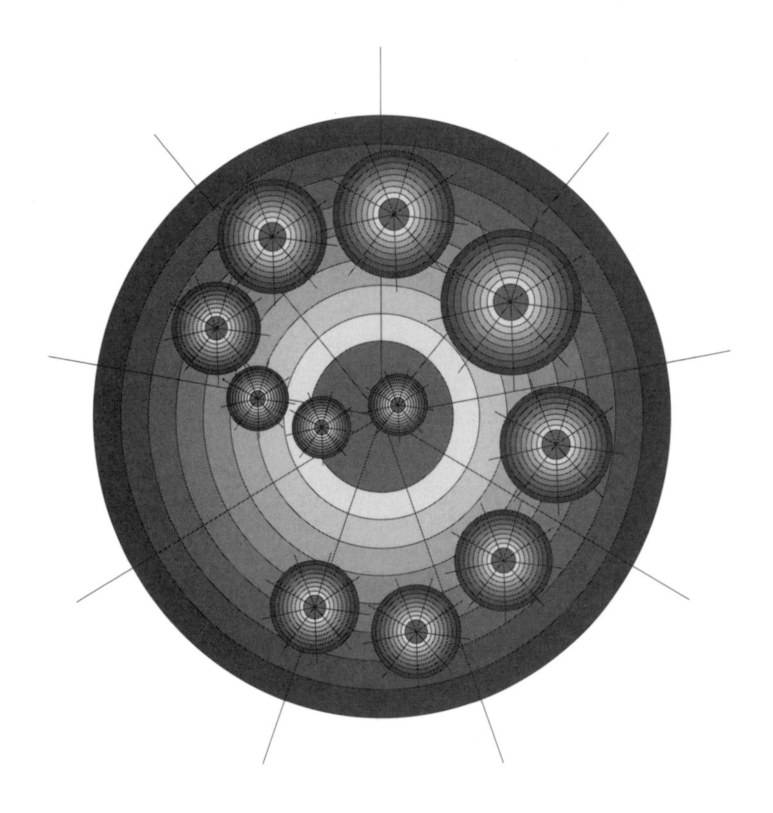

The End

In the end the numbers are all held by nine. It is as if in the beginning there is a great stillness. Only the black hole of nine reverberates.

Then the numbers slip out, first One and Eight, then Two and Seven, Three and Six, and finally Four and Five, in pairs, to take their place around the sigma circle.

The farthest away are given the most movement, to cut and dance across the inner space of the circle, as we saw in the shape of multiplications. The other numbers have less movement as they come nearer to Nine: One and Eight just move around the circumference of the circle. And the ninth spot remains unmoving.

In this secret world of arithmetic, nine controls the other numbers, releasing them into the world yet holding onto them tightly. And the sigma circle is its crucible into which all secret arithmetic flows, imprinted by a hidden code. The beauty of NINE is that it is the Alpha and the Omega of these fabrications, an organising power of vanishing and emergence.

Nine is the centre and binding rim of the prayer wheel of numbers.

And the last movement of nine never seems to come, each revelation or discovery simply deepens the mystery. The fascination grows. Like a spiral the shape of nine continues to evade a simple end, winding itself further into enigma and exploration. Enjil said that the Mandala and his quest for nine was but a reflection on life: Who is the man or woman, he asked, who would not like to know the hidden path that holds on to all movement? Was he not right?

In the labyrinth of appearances with all its shouting, twists and turns, most of us become lost and bewildered. To find our way we need a code. On the surfaces of bent experience the straightness of our logic is not enough – there are no clues to a deeper understanding, no whispers that we must hear to make our inner world hold strong and have meaning.

At the heart of the story of Enjil and the Mandalas is the simple truth, that a secret in itself is beautiful and once that is known, then somehow the fact gains power and multiplies. The world that grows around it is never barren or wasted, for in every part we see the trace of the original idea. The many that is one has always been the greatest treasure to find.

In the eternal abstraction of points, number 9, will always find connections. To those who know how to look, the insights will grow.

There is no end, as long as there are the numbers.

Acknowledgments

A book starts out as a solo journey but soon the help of others is needed and a support team builds up whose advice and help are fundamental to completing the mission. No one's help was more invaluable than Margaret Cashin's who typed and edited the manuscript and advised on layout plus helped research original material.

I thank also Peter Speleers for his patience and skill in helping with the computer drawings and transforming rough concept scrawls into beautiful finished images.

The Author

Cecil Balmond was born in Sri Lanka and is a designer and structural engineer of world renown. He is Deputy Chairman of Ove Arup and Partners Ltd. and leads a design unit of scientists, architects and engineers to pursue his interest in the genesis of form using numbers, music and mathematics as vital sources.

Balmond is the author of *informal* (2002) and *Element* (2007) both by Prestel. He has received many awards for his work including the Banister Fletcher Prize for *informal* as best book on architecture, the Gengo Matsui Prize and the RIBA Charles Jencks Award for Theory in Practice. He has taught at Harvard GSD and at the Yale School of Architecture and he now holds the Crét Chair at PennDesign as Professor of Architecture.

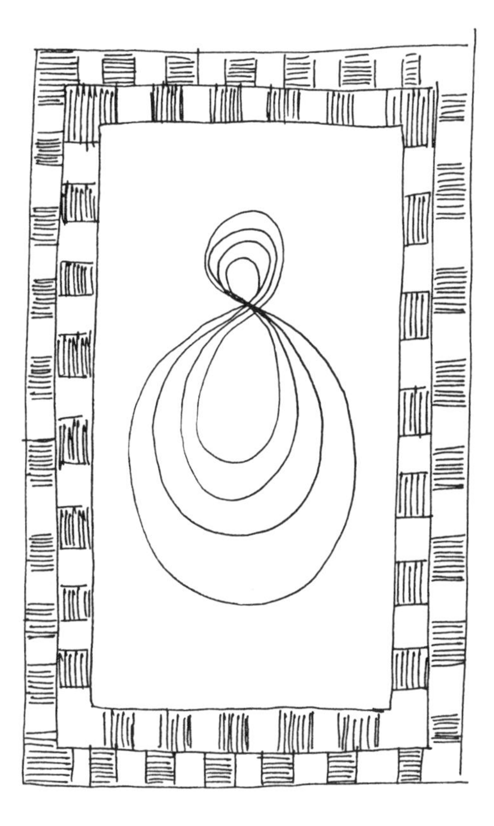

Selected Bibliography

The books I read, studied and enjoyed:

A Mathematicians Apology, – G. H. Hardy, (1940) Cambridge University Press – a lovely soliloquy on mathematics, the art and creativity. Hardy describes well the joy of discovering something in mathematics, just for itself, an invention being held only by its internal logic.

Recreations in the Theory of Numbers, Albert H. Beiler, (1964) Dover Publications Inc., N.Y. – a treasure trove on the manipulation of numbers with mathematical puzzles and several problems set for the keen reader to solve. The number pattern on page 167 in my book is obtained from Beilers's book.

Excursions in Number Theory, C. Stanley Ogilvey & John T., unabridged, corrected (1988) republication (originally published 1964) Dover Publications Inc., N.Y. – a well written book on key aspects of numbers, primes, complex numbers, continued fractions, etc. easy to follow for someone with higher school maths.

Figuring: The Joy of Numbers, Shakuntala Devi, (1977), Penguin (1990) – a simple highly enjoyable book on numbers for the lay reader on arithmetic process; plus tricks and puzzles and easy to follow. Shakuntala Devi came to my school once, doing huge calculations in her head at great speed. She is billed as the world's fastest human computer.

Numbers – Old and New, Irving and Ruth Adler, (1960) Dennis Dobson – a small book for children on numbers and their histories, with clear illustrations by Peggy Adler to describe the concepts.

The Magic of Numbers, Eric Temple Bell, (1991) Dover Publications Inc., N.Y. – a celebration of the spirit of Pythagoras; an interesting read on the nature and meaning of numbers from ancient times.

Multicultural Mathematics, David Nelson, George Gheverghese Joseph and Julian Williams, (1963) Oxford University Press – a book on the teaching of mathematics from the background of several cultures. Interesting and stimulating. I found the ancient Chinese method of counting up to 99,999 on one hand here.

The Divine Proportion, H.E. Huntley, (1970) Dover Publications Inc., N.Y. – comprehensive book on the golden ratio and its beauty in art, mathematics and nature.

Numerology for the New Age, Lynn M. Buess, (1980) De Vorss & Co – a modern interpretation on the old science of how numbers are used to divine character and personality. A very entertaining read for those who follow numbers beyond the literal.

Nine Star Ki, Bob Sachs, (1996) Element Books – a full explanation of the Chinese astrological system based on 1 to 9.

The Masks of God, Joseph Campbell, (1973) Souvenir Press (British Edition) – a huge and great book about the psychology of myth and ritual. A classic work that inspired me as a young man to trust in 'magic'.

Man and his Symbols, Carl Jung, (1964) Aldus Books, Picador (1978) – a work on the meaning of symbol and archetype. I was interested in the reference to a Yantra (or Mandala) of nine linked triangles.

I browsed through the following dictionaries and encyclopaedias of mythologies:

The Wordsworth Dictionary of Mythology, (1994) Wordsworth.

Brewers Dictionary of Phrase and Fable, (1959) 15th edition 1995, Cassel Publishers Ltd.

Essential Teachings of Islam, Kerry Brown, (1990) Arrow Books.

Man, Myth and Magic: The Illustrated Encyclopaedia of Mythology, Religion and the Unknown, Richard Cavendish (ed.) (1983) Marshall Cavendish.

Penguin Dictionary of Religions, John Hinnells (ed.) (1986) Penguin Books.

And

The Bible.

… Σ the end + 1 + 1 = ?